T0301760

Scientific Basis for Nuclear Waste Management XXXVIII

MATERIALS RESEARCH SOCIETY
SYMPOSIUM PROCEEDINGS VOLUME 1744

Scientific Basis for Nuclear Waste Management XXXVIII

Symposium held November 30-December 5, 2014, Boston, Massachusetts, U.S.A.

EDITORS

Josef Matyáš
Pacific Northwest National Laboratory
Richland, WA

Stéphane Gin
CEA
Bagnols-sur-Ceze, France

Robert Jubin
Oak Ridge National Laboratory
Oak Ridge, TN

Eric Vance
ANSTO
Kirrawee, Australia

Materials Research Society
Warrendale, Pennsylvania

CAMBRIDGE
UNIVERSITY PRESS

Shaftesbury Road, Cambridge CB2 8EA, United Kingdom

One Liberty Plaza, 20th Floor, New York, NY 10006, USA

477 Williamstown Road, Port Melbourne, VIC 3207, Australia

314–321, 3rd Floor, Plot 3, Splendor Forum, Jasola District Centre, New Delhi – 110025, India

103 Penang Road, #05–06/07, Visioncrest Commercial, Singapore 238467

Cambridge University Press is part of Cambridge University Press & Assessment, a department of the University of Cambridge.

We share the University's mission to contribute to society through the pursuit of education, learning and research at the highest international levels of excellence.

www.cambridge.org
Information on this title: www.cambridge.org/9781605117218

Materials Research Society
506 Keystone Drive, Warrendale, PA 15086
http://www.mrs.org

First published 2015

CODEN: MRSPDH

A catalogue record for this publication is available from the British Library

ISBN   978-1-605-11721-8   Hardback

CONTENTS

CAPTURE AND IMMOBILIZATION OF RADIONUCLIDES

*Invited Paper

DEVELOPMENT AND CHARACTERIZATION OF WASTE FORMS

*Invited Paper

*Invited Paper

STORAGE AND DISPOSAL OF NUCLEAR WASTE

*Invited Paper

PREFACE

The Materials Research Society's Symposium EE, entitled "Scientific Basis for Nuclear Waste Management XXXVIII," was held November 30th through December 5th, 2014, at the MRS Fall Meeting in Boston, Massachusetts. The symposium discussed the key scientific challenges for the safe and effective management of spent nuclear fuel and radioactive waste and provided an overview of the international research and waste management programs around the world. Waste forms and engineered barrier system properties, interactions between engineered and geological systems, radiation effects, chemistry and transport of radionuclides, and long-term predictions of repository performance were just some of the topics presented at the symposium by internationally renowned speakers and leading researchers in the field.

The symposium attracted 85 abstracts. This proceedings volume contains 31 papers from the meeting. The papers were divided into four sections: (1) Capture and Immobilization of Radionuclides, (2) Development and Characterization of Waste Forms, (3) Corrosion Behavior of Materials, and (4) Storage and Disposal of Nuclear Waste. Each paper provides a glimpse of the recent advances in nuclear waste management, which presents a global challenge for further development of the nuclear power industry. In spite of significant opposition worldwide after the accident at the Fukushima Daiichi nuclear power plant, we hope that over the next few decades, current and future generations of scientists and technologists will design, implement, and communicate an integrated understanding of the multi-scale processes involved in the processing, packaging, disposal, and regulation of the wide variety of materials designated as nuclear waste.

Josef Matyáš
Stéphane Gin
Robert Jubin
Eric Vance

June 2015

Acknowledgments

The organizers of this symposium sincerely thank all of the oral and poster presenters who contributed to this proceedings volume. We also thank the reviewers for valuable feedback to the editors and to the authors. The organizers of Symposium are grateful to the United States Department of Energy Office of River Protection for its financial support.

MATERIALS RESEARCH SOCIETY SYMPOSIUM PROCEEDINGS

MATERIALS RESEARCH SOCIETY SYMPOSIUM PROCEEDINGS

Prior Materials Research Symposium Proceedings available by contacting Materials Research Society

Capture and Immobilization of Radionuclides

Mater. Res. Soc. Symp. Proc. Vol. 1744 © 2015 Materials Research Society
DOI: 10.1557/opl.2015.297

Current Status of Immobilization Techniques for Geological Disposal of Radioactive Iodine in Japan

Kazuya Idemitsu [1], Tomofumi Sakuragi [2]
[1] Department of Applied Quantum Physics and Nuclear Engineering, Kyushu University, Fukuoka, Japan
[2] Radioactive Waste Management Funding and Research Center, Tokyo, Japan

ABSTRACT

Nuclear reprocessing plants in Japan produce radioactive iodine-bearing materials such as spent silver adsorbents. Japanese disposal plans classify radioactive waste containing a given quantity of iodine-129 as Transuranic Waste Group 1 for spent silver adsorbent or as Group 3 for bitumen-solidified waste, and stipulate that such waste must be disposed of by burial deep underground. Given the long half-life of iodine-129 of 15.7 million years, it is difficult to prevent release of iodine-129 from the waste into the surrounding environment in the long term. Moreover, because ionic iodine is soluble and not readily adsorbed, its migration is not significantly retarded by engineered or natural barriers. The release of iodine-129 from nuclear waste therefore must be restricted to permit reliable safety assessment; this technique is called "controlled release". It is desirable that the release period for iodine be longer than 100,000 years. To this end, several techniques for immobilization of iodine have been developed; three leading techniques are the use of synthetic rock (alumina matrix solidification), BPI ($BiPbO_2I$) glass, and high-performance cement. Iodine is fixed as AgI in the grain boundary of corundum or quartz through hot isostatic pressing in synthetic rock, as BPI in boron/lead-based glass, or as cement minerals such as ettringite in high-performance alumina cement. These techniques are assessed by three models: the corrosion model, the leaching model, and the solubility-equilibrium model. This paper describes the current status of these three techniques.

INTRODUCTION

In Japanese nuclear reprocessing plants, radioactive iodine in the off-gas pretreatment process is collected using silver adsorbent. This spent silver adsorbent includes [129]I and is categorized as Transuranic (TRU) Waste Group 1 for geological disposal [1]. In the case of the Rokkasho Reprocessing plant in northern Japan, which has a maximum reprocessing capacity of 800 tU/year, 40 years of operation is estimated to produce 51 TBq of [129]I. The silver-sorbent used for iodine filters is an alumina-based material loaded with silver nitrate ($AgNO_3$). Iodine reacts with $AgNO_3$ and is fixed as silver iodide (AgI), or silver iodate ($AgIO_3$) like in other silver-doped sorbents used for iodine [2, 3]. Trapping of iodine by silver-loaded sorbents is expected to follow the reaction

$$3AgNO_3 + \frac{3}{2}I_2 = 2AgI + AgIO_3 + 3NO_2. \tag{1}$$

Because the half-life of [129]I is 15.7 million years, it is difficult to prevent the release of [129]I from waste into the surrounding environment in the long term. Moreover, as [129]I exists in

3

groundwater in anionic form, it is not expected that ^{129}I dispersal will be retarded by sorption onto most of the barrier materials used in the geological disposal environment. Thus, the peak exposure dose is easily affected by geological environmental conditions such as groundwater flow rate. To lower the peak ^{129}I exposure dose, it is more effective to decrease the rate of dissolution of waste or to elongate the release period, rather than to delay the start of dissolution. This approach is called "controlled release". To obtain the desired iodine release-delaying capability for waste, the relationship between the maximum ^{129}I exposure dose and the iodine release period has been assessed through calculations [4]. Results show that the peak exposure dose from ^{129}I in each hydrologic condition assessed could be made lower than 10 μSv/year when the iodine release period was longer than 10^5 years, which has led to the goal of developing waste forms that can suppress release for at least this duration.

Several techniques for immobilization of iodine have been developed for this purpose [5-12]. These have been narrowed down to three techniques: synthetic rock (alumina matrix solidification), BPI (BiPbO$_2$I) glass, and high-performance cement. These techniques were classified into three types according to the model used for their assessment, namely, the leaching model, the distribution equilibrium model, and the solubility-equilibrium model. In this paper, the current status of these techniques is described.

SYNTHETIC ROCK (ALUMINA MATRIX SOLIDIFICATION) BY THE HOT ISOSTATIC PRESS PROCESS

In synthetic rock, iodine in the form of silver iodide is sealed within the grain boundaries of densely solidified corundum through hot isostatic pressing (HIP). Because natural hard rock provides long-term durability with a matrix that is difficult to dissolve and that does not easily allow water to penetrate, the contained chemical elements are not easily released. Therefore, it is assumed that rocks converted to a crystalline material by HIP are stable and capable of immobilizing iodine over long periods.

In HIP process, the waste silver adsorbent is pulverized and packed into a capsule, which is pre-heated at 450 °C in vacuum for 2 h, then heated at 1200 °C and 175 MPa for 3 h as shown in Fig. 1. This procedure generates very little secondary waste.

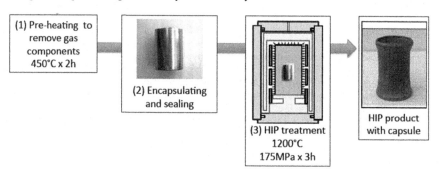

Fig. 1 HIP process for immobilization of iodine fixed in Ag-alumina filter

4

During pre-heating, silver iodate and silver nitrate are decomposed to silver iodide and silver as shown by the following reactions

$$AgIO_3 = AgI + \frac{3}{2}O_2, \qquad\qquad\qquad\qquad (2)$$

$$AgNO_3 = Ag + NO_2 + \frac{1}{2}O_2. \qquad\qquad\qquad (3)$$

Iodine is not released as a gas [13]. The microstructure of the HIP product is shown in Fig. 2. Silver iodide is surrounded by alumina. The concentration of AgI in the waste form was approximately 20wt%. A certain amount of void (black parts in Fig. 2) was observed. The void ratio could be reduced to less than 5% by HIP after pre-heating at 450 °C for 3 h in vacuum below 7×10^{-2} Pa, whereas around 15% porosity remained with no pre-heating, as shown in Fig. 3. Uniaxial compressive strength observed for the HIP product was as high as 1.1 GPa.

Fig. 2 Microstructure of HIP product

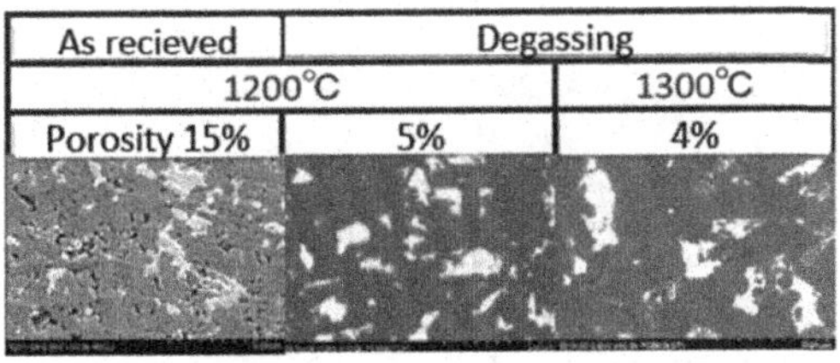

As recieved	Degassing	
1200°C		1300°C
Porosity 15%	5%	4%

Fig. 3 Porosity reduction by degassing at 450 °C for 3 h in vacuum before HIP

Silver iodide is among the most stable iodides and has an extremely low solubility product, as low as 10^{-16} at 25 °C [14]. Although silver might be reduced to metallic silver, which could lead to a release of iodine under reducing conditions, a metallic silver layer could form and protect against the transport of reactants, which could subsequently suppress the release of iodine [15]. However, hydrosulfide concentrations as high as 1 mM could increase the discharge of iodine. X-ray diffraction patterns of HIP products are shown in Fig. 4 before and after contact with leachant containing 1 mM of Na_2S at 25 and 60 °C. After contact with leachant, silver sulfide was observed in the HIP product, as shown in Fig. 5. It is assumed that the following reaction occurs

$$2AgI + HS^- = Ag_2S + 2I^- + H^+. \tag{4}$$

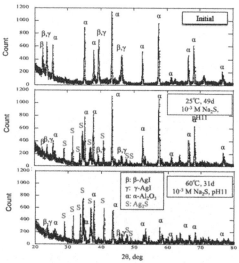

Fig. 4 XRD patterns of HIP product before and after contact with leachant containing 1 mM Na_2S at 25 and 60 °C

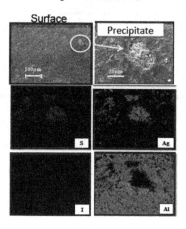

6

Fig. 5 Silver sulfide precipitation by contact with leachant containing 1 mM Na₂S

Because silver sulfide was precipitated in the HIP product, hydrosulfides could diffuse into the HIP product through pores or grain boundaries. Therefore, it is necessary to avoid exposure to hydrosulfide to suppress discharge of iodine from the HIP product.

BPI GLASS

An inorganic anion exchanger, $BiPbO_2NO_3$, was developed to remove and immobilize various kinds of industrial waste liquids and anions, in particular, halogens generated as a result of nuclear power generation [16, 17]. The vitrification process of BPI glass is shown in Fig. 6. Initially, H_2 gas is used as a reductant to release iodine from a spent iodine filter. Then, the iodide ion is included in the BPI through a reaction with an inorganic anion exchanger, $BiPbO_2NO_3$. The ion exchange reaction is as follows,

$$BiPbO_2NO_3 + I^- = BiPbO_2I + NO_3^-. \quad (5)$$

The BPI was mixed with a glass frit, the composition of which is also shown in Fig. 6, and was heated to 540 °C for vitrification. A low melting temperature of 540 °C is used to avoid iodine volatilization during vitrification. The target composition of the BPI vitrified waste form is shown in Table I. The BPI glass is formed by mixing BPI with B_2O_3/PbO-based glass, a type of lead-containing glass used for radiation shielding. The crystal structure of BPI ($BiPbO_2I$) is chemically stable and is similar to that of the natural mineral perite ($BiPbO_2Cl$), as is shown in Fig. 6.

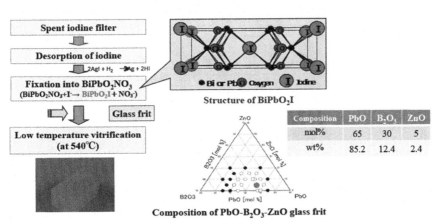

Composition of PbO-B_2O_3-ZnO glass frit

Fig. 6 Vitrification process and structure of BPI glass

Table I. Target composition of the BPI glass

Elements	I	B	Pb	Zn	Bi	O
mol%	1.0	19.8	22.1	1.6	1.0	54.5

Leaching examinations were conducted by using the leachants shown in Table II. The result of the leaching examinations is shown in Fig. 7, where normalized mass loss of iodine is plotted as a function of normalized mass loss of boron. As shown in Fig. 7, iodine and boron dissolved congruently and the amount of dissolution increased with increasing concentration of bicarbonate. Solids were observed at the end of leaching experiments and altered layers and precipitates were found (see Fig. 8). The thicknesses of altered layers in the experiments using synthetic seawater and equilibrated with bentonite were plotted as a function of leaching period. The precipitates were found to be $Pb_3(CO_3)_2(OH)_2$, hydrocerussite, $NaPb_2(CO_3)_2OH$, $CaCO_3$, and aragonite. It is therefore necessary to avoid carbonates to suppress discharge of iodine from BPI glass.

Table II Conditions of leaching experiments running for up to 800 days

Leachant	Reducing agent	S/V (cm^{-1})
NaCl 0.0055, 0.055, 0.55 M	Electrolytic iron	0.1
Synthetic sea water NaCl 0.55 M + NaHCO$_3$ 0.05 M	Electrolytic iron	0.1
Equilibrated water with bentonite (Kunigel-V1)	Electrolytic iron	0.1 - 9.0
Equilibrated water with cement (OPC, HFSC)	Electrolytic iron	0.1

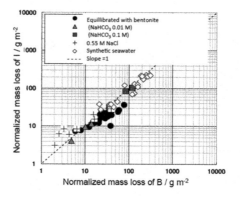

Fig. 7 Relation between normalized mass loss of iodine and boron

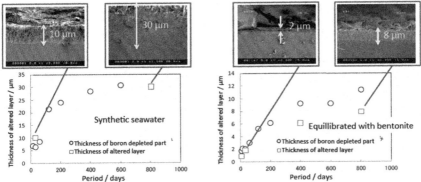

Fig. 8 The thickness of altered layers observed on BPI glass contact with synthetic seawater and equilibrated with bentonite as a function of the leaching period

HIGH-PERFORMANCE CEMENT

Generally, iodine weakly adsorbs onto ordinary cementitious materials. However, some cement hydrates such as ettringite (AFt: $Ca_6[Al(OH)_6]_2 (SO_4)_3 \cdot 26H_2O$) and monosulfate (AFm: $[Ca_2Al(OH)_6]_2(SO_4) \cdot 6H_2O$) easily adsorb anions. The high-performance cement technique uses sulfate-added calcium aluminate cement. Iodine is fixed as IO_3^- due to its high affinity for AFt and AFm [12]. The process for the immobilization of iodine in cement is shown in Fig. 9. Iodine is separated from spent silver filters in alkaline solution by treatment with Na_2S and is then converted into iodate ions via treatment with ozone. The iodate ions are immobilized through kneading and solidification with alumina cement and gypsum in a weight ratio of 100:15.5.

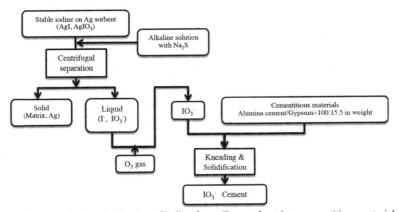

Fig. 9 Process for immobilization of iodine from silver sorbent into cementitious material

9

Because of the pretreatment in alkaline solutions, phase transition of iodine into a gas is rare, and the iodine recovery rate is extremely high at over 99.96%. The iodine content in the cement form was estimated to be about 2%.

Immersion experiments have been conducted by using simulated ground water and seawater to investigate the iodine release behavior from the iodine-immobilized cement under geological disposal conditions [18]. Iodine-immobilized cement for immersion experiment was prepared by mixing 0.4 M of $NaIO_3$ solution and calcium aluminate cement including gypsum. The cement was ground and grains 250 to 355 μm in diameter were collected as the sample. Fresh groundwater (FGW) and saline groundwater (SGW) were prepared for the simulated groundwater used as immersion solution. Powdered sample and immersion solution were placed into a container at a given liquid-solid ratio and kept in a glove box under inert atmosphere. The iodine release fraction as a function of the liquid-solid ratio is plotted in Fig. 10. Higher salinity enhanced the release of iodine. Iodine release could also be caused by dissolution of minerals in cement waste. Mineral composition and iodine distribution on the minerals is shown in Fig. 11 and Fig. 12, respectively. Initially, iodine existed in AFt, AFm, and hydrogarnet (HG). The result of mineral composition analysis indicated that AFt and AFm decreased with the increase in the liquid-solid ratio, whereas calcite increased. Most iodine existed in the AFt phase, was released mainly with the dissolution of AFt. Monosulfate dissolved out at a lower liquid-solid ratio than did AFt. In SGW conditions, almost all iodine was released at liquid-solid ratios between 50 and 150 cm³/g and in FGW conditions at liquid-solid ratios between 600 and 900 cm³/g. Calculation of the iodine release fraction was conducted using the thermodynamic data in Table III [19] and is also shown in Fig. 10. Carbonate in the immersion solution could also affect the dissolution of iodine-bearing minerals such as AFt, since calcium would be consumed with carbonate.

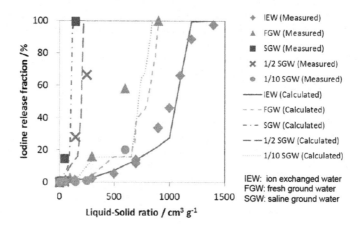

Fig. 10 Iodine release fraction as a function of liquid/solid ratio

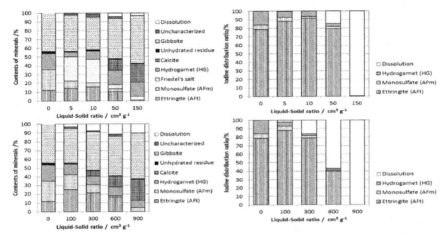

Fig. 11 Relationship between mineral
composition and liquid/solid ratio
(top, SGW; bottom, FGW)

Fig. 12 Relationship between iodine distribution
on minerals and liquid/solid ratio
(top, SGW; bottom, FGW)

Table III Thermodynamic data of iodine-bearing minerals at 25 °C [19]

Minerals	Chemical reaction	log K
IS-AFt	$Ca_6(Al(OH)_6)_2 \cdot 3((IO_3)_{0.5}(SO_4)_{0.5}) \cdot 24.5(H_2O) \cdot (OH)_{1.5}$ $= 6Ca^{2+} + 2Al(OH)_4^- + 1.5IO_3^- + 1.5SO_4^{2-} + 5.5OH^- + 24.5H_2O$	-51.29
AFt	$Ca_6Al_2O_6 \cdot (SO_4)_3{:}32H_2O$ $= 6Ca^{2+} + 2Al(OH)_4^- + 3SO_4^{2-} + 4OH^- + 26H_2O$	-43.94
IO_3-AFm	$(CaO)_3 \cdot Al_2O_3 \cdot Ca(IO_3)_2{:}12H_2O$ $= 4Ca^{2+} + 2Al(OH)_4^- + 2IO_3^- + 4OH^- + 6H_2O$	-36.8
I-HG	$Ca_3Al_2(OH)_8(IO_3)_4 = 3Ca^{2+} + 2Al(OH)_4^- + 4IO_3^-$	- 33.20

SUMMARY

The peak exposure dose from [129]I in each geologic condition observed could be reduced
to lower than 10 µSv/year when the iodine release period was longer than 10^5 years.
Consequently, the goal is to develop waste forms that can suppress release for at least 100,000
years. Three techniques have been developed for this purpose, synthetic rock (alumina matrix
solidification) by HIP, BPI glass, and high-performance alumina cement.

For synthetic rock, hydrosulfide as high as 1 mM could affect iodine release due to silver
sulfide precipitation. Hydrosulfide could diffuse through pores or the grain boundary of the
corundum matrix. Pre-heating for degassing before HIP is effective in reducing porosity. For

11

BPI glass, iodine and boron could dissolve congruently and carbonate affected iodine release due to hydrocerussite (Pb$_3$(CO$_3$)$_2$(OH)$_2$) precipitation. For high-performance alumina cement, the solubility of iodine-bearing minerals such as Aft, Afm, and HG could control iodine release and salinity, and carbonate enhanced the release of iodine.

In the future, it will be necessary to acquire process data in order to scale up the process and further understand the iodine release mechanism.

ACKNOWLEDGMENTS

These studies represent part of the research and development of the processing and disposal techniques for TRU waste containing ^{129}I and C-14 (FY 2013) program "Research and Development for TRU Waste Disposal in Japan", funded by the Agency of Natural Resources and Energy, Ministry of Economy, Trade and Industry of Japan.

REFERENCES

[1] Federation of Electric Power Companies (FEPC) and Japan Atomic Energy Agency (JAEA), "Second Progress Report on Research and Development for TRU Waste Disposal in Japan," Repository Design, Safety Assessment Means of Implementation in the Generic Phase. FEPC TRU-TR2-2007-01 or JAEA Review 2007-010.

[2] M. Kikuchi, M. Kitamura, H. Yusa, S. Horiuchi, Removal of radioactive methyl iodide by silver impregnated alumina and zeolite, Nuclear Engineering and Design, **47(2)**, 283-287 (1978).

[3] V.S. Ramos, V. R. Crispim, L. E. B. Brandão, New filter for iodine applied in nuclear medicine services, Applied Radiation and Isotopes, **82**, 111-118 (2013).

[4] T. Nishimura, T. Sakuragi, Y. Nasu, H. Asano, and H. Tanabe, Development of Immobilization Techniques for Radioactive Iodine for Geological Disposal, In: *Proceedings of Workshop "Mobile Fission and Activation Products in Nuclear Waste Disposal"*, Jan. 16th–19th 2007, La Baule, France, NEA No. 6310, 221–34 (2009).

[5] H. Fujihara, T. Murae, R. Wada, and T. Nishimura, "Fixation of radioactive iodine by hot isostatic pressing", in ICEM '99: (the 7th International Conference Proceedings on Radioactive Waste Management and Environmental Remediation, Nagoya, Japan, September 26-30, 1999), pp 1-5.

[6] R. Wada, T. Nishimura, Y. Kurimoto, and T. Imakita, HIP Rock Solidification Technology for Radioactive Iodine-Contaminated Waste: *Kobe Seiko Giho* 53, 47–55 (2003); in Japanese.

[7] H. Fujihara, T. Murase, T. Nishi, K. Noshita, T. Yoshida, and M. Matsuda, Low Temperature Vitrification of Radioiodine Using AgI–Ag$_2$O–P$_2$O$_5$ Glass System, *Mater. Res. Soc. Symp. Proc.*, **556**, 375–82 (1999).

[8] H. Kato, O. Kato, and H. Tanabe, Review of Immobilization Techniques of Radioactive Iodine for Geological Disposal, In: *Proceedings of the International Symposium NUCEF-2001*, 31st Oct.–2nd Nov 2001, JAERI, Tokai, Japan; 697–704 (2002).

[9] T. Advocat, C. Fillet, F. Bart *et al.*, New Conditionings for Enhanced Separation Long-lived Radionuclides, In: *Proceedings of the International Conference "Back-End of Fuel Cycle: From Research to Solution", Global'01*, Sept. 9th–13th, 2001, Paris, France.

[10] J. Izumi, I. Yanagisawa, K. Katsurai, *et al.*, Multi-layered Distributed Waste-Form of Iodine-129: Study on Iodine Fixation of Iodine Adsorbed Zeolite by Silica CVD, In: *Proceedings of the Symposium on Waste Management 2000*, Feb. 27th–March 2nd, 2000, Tucson, Arizona, USA.

[11] T. Nakazawa, H. Kato, K. Okada, S. Ueta, and M. Mihara, Iodine Immobilization by Sodalite Waste Form, *Mater. Res. Soc. Symp. Proc.*, **663**, 51–7 (2001).

[12] M. Toyohara, M. Kaneko, H. Ueda, N. Mitsutsuka, H. Fujiwara, T. Murase, and N. Saito, Iodine Sorption onto Mixed Solid Alumina Cement and Calcium Compounds, *J. Nucl. Sci. Technol.*, **37**, 970–978 (2000).

[13] K. Masuda, O. Kato, Y. Tanaka, S. Nakajima, T. Sakuragi, S. Yoshida, Development of Solidification Process of Spent Silver-Sorbent using Hot Isostatic Press Technique for Iodine Immobilization, Progress in Nuclear Energy Special issue for Scientific Basis of Nuclear Fuel Cycle II 2014 (to be published).

[14] W. M. Latimer, The Oxidation States of Elements and their Potentials in Aqueous Solutions, Prentice Hall (1952).

[15] Y. Inagaki, T. Imamura, K. Idemitsu, T. Arima, A. Kato, T. Nishimura, H. Asano, Aqueous Dissolution of Silver Iodide and Associated Iodine Release under Reducing Conditions with FeCl$_2$ Solution, J. Nucl. Sci. Technol., **45(9)**, 859-866 (2008).

[16] H. Kodama, A. Dyer, M. J. Hudson, P. A. Williams, "Progress in Ion Exchange," The Royal Society of Chemistry, Cambridge, UK, p. 39 (1997)

[17] H. Kodama, N. Kabay, "Reactivity of Inorganic Anion Exchanger BiPbO$_2$NO$_3$ with Fluoride Ions in Solution," Solid State Ionics, 141–142 (2001).

[18] Y. Yamashita, S. Higuchi, M. Kaneko, R. Takahashi, T. Sakuragi, H. Owada, Iodine Release Behavior from Iodine-Immobilized Cement Solid under Geological Disposal Conditions, Progress in Nuclear Energy Special issue for Scientific Basis of Nuclear Fuel Cycle II 2014 (to be published).

[19] Y. Haruguchi, S. Higuchi, M. Obata, T. Sakuragi, R. Takahashi, H. Owada, The Study on Iodine Release Behavior from Iodine-Immobilized Cement Solid, *Mater. Res. Soc. Symp. Proc.*, **1518**, 85-90 (2013).

Mater. Res. Soc. Symp. Proc. Vol. 1744 © 2015 Materials Research Society
DOI: 10.1557/opl.2015.309

French Studies on the Development of Potential Conditioning Matrices for Iodine 129

Lionel Campayo[1], Fabienne Audubert[2], Jean-Eric Lartigue[3], Eglantine Courtois-Manara[4], Sophie Le Gallet[5], Frédéric Bernard[5], Thomas Lemesle[1], François O. Mear[6], Lionel Montagne[6], Antoine Coulon[1], Danielle Laurencin[7], Agnès Grandjean[1] and Sylvie Rossignol[8]

[1]CEA, DEN, DTCD, Bagnols-sur-Cèze, F-30207, France.
[2]CEA, DEN, DEC, Saint Paul Lez Durance, F-13108, France.
[3]CEA, DEN, DTN, Saint Paul Lez Durance, F-13108, France.
[4]Electron Microscopy Spectroscopy Laboratory, Karlsruhe Institute of Technology, P.O. Box 3640, Karlsruhe, D-76021, Germany.
[5]LICB, UMR-6303, 9 av. A. Savary, Dijon, BP 47870, F-21078, France.
[6]UCCS, UMR-8181, Université Lille 1, Villeneuve d'Ascq, F-59655, France.
[7]ICGM, UMR-5253, Université de Montpellier 2, Place Eugène Bataillon, CC1701, Montpellier, F-34095, France.
[8]GEMH, ENSCI, Centre Européen de la Céramique, 12 rue Atlantis, Limoges, F-87068, France.

ABSTRACT

Since 1991, in France, studies on the conditioning of iodine were carried out to assess the potential of several specific inorganic host matrices. The apatite family has been mainly studied because of its good chemical durability and its ability to confine iodine over geological time scales. A lead-bearing apatite, $Pb_{10}(VO_4)_{4.8}(PO_4)_{1.2}I_2$, and a calcium-bearing apatite, $Ca_{10}(PO_4)_6(OH)_{2-x}(IO_3)_x$, were selected on the basis of their incorporation rate (between 7 and 10 wt.%) and a satisfactory resistance to leaching (V_0(50 °C, pure water) ~ 10^{-2} g.m^{-2}.d^{-1}; V_r(50 °C, pure water) < 10^{-4} g.m^{-2}.d^{-1}). However, with such materials, the removal of open porosity requires non conventional sintering techniques like spark plasma sintering to decrease the surface exposed to water. This is why, in parallel, other matrices, like silver phosphate glasses, have also been investigated. To improve the chemical durability and thermal properties of these glasses, cross-linking reagents were added to their formulation.

INTRODUCTION

Iodine management has received a special attention since the development of nuclear industry [1]. Actually, this element is thought to be one of the main contributors to the dose released in the biosphere by a disposal site [2]. This can be explained by poor retention properties of geological formations generally considered for disposal sites. Consequently, in the case of an iodine management that would rely on its conditioning, the chemical durability of the conditioning matrix is of primary importance. Even if no criteria currently exist, matrices which have a chemical durability at least as good as the one of borosilicate glasses designed for the conditioning of high level wastes (e.g., R7T7 glass) are targeted. This makes the development of iodine-bearing conditioning matrices particularly challenging. In France, several matrices aiming at conditioning iodine were studied during the past two decades. On the basis of the study of natural specimens [3], apatites have appeared as good candidates. Not only do they have incorporated iodine 129 in their crystalline structure [3], but they also show a good stability in neutral to alkaline media (the ones generally met for potential disposal sites) [4]. The question of the shaping of iodine-bearing apatites, which have a poor thermal stability, is still a concern but the emergence of non conventional sintering techniques like spark plasma sintering (SPS) has allowed partially overcoming this difficulty. Lead-bearing apatites have been the first to benefit from these technical evolutions [5]. At the same time, public acceptance issues towards lead-bearing compounds have revived studies around apatitic compositions without hazardous metallic cation (lead, cadmium …). Calcium-bearing apatites were developed as a result of these considerations [6]. These apatites have the particularity of incorporating iodine under its iodate form. They can be easily obtained by means of "room temperature" synthesis protocols. Nonetheless, because of the drawback of their complex shaping, apatites were not the only matrices to be studied. Glasses have also attracted much attention. Here, the case of silver phosphate glasses which are able to incorporate huge amounts of silver iodide has been investigated. It is worth noting that these glasses have already been proposed as potential conditioning matrices for iodine by Japanese teams in the late 90's [7,8]. The idea was here to improve their properties modifying their composition with cross-linking reagents able to create new chemical bonds between phosphate chains [9]. The present article is an illustration of the different kinds of studies that were carried out on these topics: a focus on chemical durability of lead-bearing apatite, on the sinterability of calcium-bearing apatite and on the improvement of thermal properties of silver phosphate glasses is showed hereafter.

EXPERIMENT

Details on apatite (either for lead-bearing or calcium-bearing apatite) or silver phosphate glasses preparation have already been published elsewhere [4,5,6,9].

Here, the case of the long term behavior of lead-bearing apatite is discussed through its alteration in pure water at 90 °C. A surface to volume of water ratio of 30 cm^{-1} has been considered. Iodine and lead concentrations were measured out by ICP-MS. It is worth noting that, due to the polycrystalline feature of this material, grain size fraction consists of aggregates of small crystals the size of which ranges between 1 and 8 μm. The normalized mass loss (NL, g.m^{-2}) was calculated according to the following equation:

$$NL_i(t) = \frac{[i]_t . V}{f_i . S}$$ (1)

Where $[i]_t$ is the elemental concentration for time "t" (g.L^{-1}), V is the volume of water (L), f_i is the initial weight fraction of element "i" in the solid and S is the surface of solid (m^2). The leaching rate (R, g.m^{-2}.d^{-1}) represents the derivative of the normalized mass loss as a function of time.

For sintering experiments of calcium-bearing apatite, 4.90 g of powder was SPS treated under nitrogen atmosphere (~ 1 atm) in graphite molds (inner diameter of 30.7 mm, external diameter of 90 mm, height of 50 mm) with a FCT system HPD 125 (FCT, Germany). A special care was devoted to the packaging of powders prior to sintering. Especially, the level of "hydration" of the as-synthesized apatites was controlled according the procedure described by Grossin et al. [10]. The shrinkage was collected as a function of temperature (p = 100 MPa, heating ramp of 60 °C.min^{-1}) up to 500 °C for three hydration levels which qualitatively correspond to a maximum value (powder P1), a median value (powder P2) and a completely dried powder (powder P3). P2 and P3 have close specific surface areas, around 64 m^2.g^{-1}, whereas P1 has a specific surface area of 118 m^2.g^{-1}.

Thermal properties of silver phosphate glasses were recorded as a function of the cross-linking reagent content, namely alumina, the amount of which was measured out by EPMA.

DSC experiments were performed at 10°C/min, under N_2 flow with a DSC 131 (Setaram). 20 mg of crushed sample were analyzed in an aluminum crucible.

DISCUSSION

Long-term behavior of lead-bearing apatite

It had been shown that lead-bearing apatite ($Pb_{10}(VO_4)_{4.8}(PO_4)_{1.2}I_2$) have an initial dissolution rate around 3.10^{-3} $g.m^{-2}.d^{-1}$ at 90 °C in pure water and that such a value was maintained over a pH range between 5.2 and 8.2 [4]. However, a point that was still unclear was the existence (or not) of a decrease of this rate with time. Actually, it is well known that HLW glasses, such as alumino-borosilicate glasses, exhibits such a decrease due to the formation of a "gel" at their surface. This gel layer limits elemental release in solution and is responsible for a drop of the dissolution rate by several orders of magnitude. Consequently, this property is important to ensure the confinement of long-lived radionuclides. It has also to be confirmed for lead-bearing apatite, the potential role of which would be to confine iodine 129 which has one of the longest half-life among fission products ($15.7\ 10^6$ years). Figure 1 shows the normalized mass losses for iodine and lead in pure water at 90 °C as well as the evolution of pH.

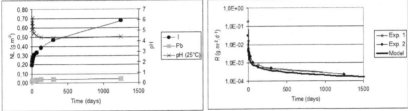

Figure 1. Normalized mass loss for lead and iodine (on the left) and leaching rate on the basis of iodine release (on the right) of a lead-bearing apatite, $Pb_{10}(VO_4)_{4.8}(PO_4)_{1.2}I_2$, altered in pure water at 90 °C (S/V = 30 cm^{-1}).

After the first year, no extra release of lead was observed. This result was explained by the precipitation of by-products containing lead, analogous to chervetite ($Pb_2V_2O_7$), the presence of which was already detected by TEM for short-term experiments [4]. As a consequence, iodine was the only possible tracer for the alteration and the behavior of the material was subsequently extrapolated from its release. Even after four years, the dissolution rate was found to decrease and the experiment was finally stopped at this duration. The final measurement gave a dissolution rate of $2\ 10^{-4}$ $g.m^{-2}.d^{-1}$. The characterization by SEM of the solid at the end of the test did not show a significant evolution by comparison with the pristine material. Noticeably, no alteration layer was observed. On the other hand, XRD characterization (results not shown here) revealed the formation of a phase of which the diffractogram is very close to that of hydroxyl-vanadinite, $Pb_{10}(VO_4)_6(OH)_2$. The combination of these results together with the evolution of the pH value during the test has led us to propose a pseudomorphic transformation of the lead-bearing iodoapatite into a lead-bearing hydroxyl-apatite (on the basis of an ionic exchange $I^- \leftrightarrow OH^-$) to explain the long-term behavior of such a matrix. This mechanism is supported by diffusion across the tunnels of the apatitic structure. Notably, the long-term dissolution rate curves of lead-bearing iodoapatite can be modeled by a diffusion process. Actually, if normalized mass loss based on iodine release is converted into thickness of the corresponding altered layer (density of lead-bearing apatite being equal to 7.117 $g.cm^{-3}$ and thickness being the ratio between normalized mass loss and density) and plotted as a function of the square root of time, a linear evolution is noted (width = $a_0 + (D*t)^{0.5}$ where t is time (s)). The slope ($D^{0.5}$) is equal to $6.71\ 10^{-12}\ m.s^{-0.5}$ (goodness of fit 0.999) which corresponds to a diffusion coefficient of $4.5\ 10^{-23}\ m^2.s^{-1}$. This value is for a leaching test performed in pure water at 90 °C for a powder of which the size fraction was comprised between 100 and 200 µm ($S_{BET(Kr)}$ = 301 $cm^2.g^{-1}$; S/V = 30 cm^{-1}).

"Sinterability" of calcium-bearing apatite

The removal of open porosity in iodoapatites allows the decrease of iodine release by contact with groundwater. It merely relies on a "mechanical" effect on the available surface able to interact with leaching media. However, sintering of these materials can be limited by their low thermal stability. This is why sintering in closed systems, like metallic canisters, by HIP were first considered to achieve relative densities above 92 % [11]. Given that the use of such devices in a nuclearised environment could be difficult, we tried to develop a liquid sintering, the kinetics of which can be accelerated by a uniaxial pressure and current in an "open" environment (i.e., sintering by SPS). This point is illustrated in Figure 2.

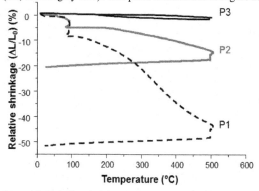

Figure 2. Relative shrinkage of a calcium-bearing iodoapatite heat-treated by SPS (p = 100 MPa, heating rate of 60 °C.min^{-1}) for three hydration ratios: maximum ratio (P1), median ratio (P2) and without hydration layer (P3).

For calcium-bearing apatites, synthesis conditions allow the formation of a hydrated layer at the surface of grains which is involved in ionic exchanges with solutions and could play a role in biological processes [10]. The stabilization of this hydrated layer strongly depends on drying conditions. As matter of fact, without this hydration layer, no shrinkage can be observed during the heat treatment by SPS. We have found that a similar behavior was observed for calcium-bearing iodoapatite here studied for the conditioning of iodine. The optimization of synthesis, drying and SPS conditions has allowed a dense pellet to be produced (near ~ 90 % of the theoretical density) at 250 °C without iodine volatilization (a slight iodine volatilization being observed for sintering temperatures equal to or above 300 °C as a function of heating rates and dwell times).

Role of alumina on thermal properties of silver phosphate glasses

Considering that the development of pressure-assisted sintering devices for nuclear industry might be complicate (either for closed or open systems), materials able to be produced by "conventional" vitrification techniques were also investigated. Silver phosphate glasses have been chosen since they can incorporate huge amounts of silver iodide [7,8]. Their properties can be influenced by their silver iodide content as well as by their Ag_2O/P_2O_5 molar ratio. Here, we try to modify these properties adding a heteroatom, the role of which would be to create new chemical bonds between phosphate units and strengthen the glassy network. For a given iodine content, namely 1 g.cm^{-3}, and a given Ag_2O/P_2O_5 molar ratio, namely 1.66, Table 1 shows the effect of the addition of alumina.

Table I. Molar composition of AgI-Ag_2O-P_2O_5-Al_2O_3 glasses and thermal properties (Glass transition temperature (T_g) and difference with crystallization onset temperature (T_x)).

Batch molar composition	Theoretical (*Measured*) Composition (mol. %)				Characteristic temperatures (°C, +/- 2)	
	AgI	Ag$_2$O	P$_2$O$_5$	Al$_2$O$_3$	T$_g$	T$_x$-T$_g$
29AgI-71 (95Ag$_5$P$_3$O$_{10}$-5Al$_2$O$_3$)	28.7 (*29.5*)	44.0 (*42.7*)	26.4 (*27.4*)	0.9 (*0.4*)	123	238
95Ag$_5$P$_3$O$_{10}$-5Al$_2$O$_3$	0.0 (*0.0*)	61.7 (*61.3*)	37.0 (*37.4*)	1.3 (*1.3*)	181	264
29AgI-71 (97Ag$_5$P$_3$O$_{10}$-3Al$_2$O$_3$)	28.8 (*29.0*)	44.2 (*44.0*)	26.5 (*26.5*)	0.5 (*0.5*)	123	257
97Ag$_5$P$_3$O$_{10}$-3Al$_2$O$_3$	0.0 (*0.0*)	62.0 (*62.4*)	37.2 (*36.7*)	0.9 (*0.8*)	171	269
29AgI-71 Ag$_5$P$_3$O$_{10}$	28.7 (*29.0*)	44.6 (*44.2*)	26.7 (*26.7*)	0.0 (*0.1*)	118	267
Ag$_5$P$_3$O$_{10}$	0.0 (*0.0*)	62.5 (*62.5*)	37.5 (*37.4*)	0.0 (*0.1*)	158	279

For these conditions (and considering an elaboration performed at 650 °C followed by a rapid quenching on a cooled brass plate), it can be noted that alumina solubility is strongly limited in the presence of silver iodide. No more than 0.4 – 0.5 mol. % can be incorporated in the glass. The remaining fraction is found in crystallizations in which aluminum is associated to phosphorus, mainly as $AlPO_4$. In spite of this poor solubility, alumina was found to create new chemical bonds between phosphate units (see reference 9 for a detailed structural analysis). As a consequence, a slight increase of T_g, from 118 to 123 °C, can be observed. This effect is more accentuated in the absence of silver iodide where alumina fully contributes to the strengthening of the glassy network. Further studies will allow the influence of alumina on chemical durability to be assessed.

CONCLUSIONS

In the frame of an iodine management relying on its conditioning, chemical durability of the conditioning matrix is of primary importance. First, an "intrinsically" good chemical durability is required. This property is controlled by the chemical composition and the structure of the material. Then, for a given matrix (i.e., for a given composition and structure), the wasteform should also display a minimum surface to be exposed to groundwater. Due to the long half-life of iodine 129, chemical durability should be precised for short-term evolution as well as for long-term evolution if a difference in the leaching mechanism occurs with time. We tried to apply this methodology to different potential conditioning matrices. Lead-bearing iodoapatites have an interesting intrinsic chemical durability. A pseudomorphic transformation of such a matrix into a lead-bearing hydroxyl-apatite (on the basis of an ionic exchange $I^- \leftrightarrow OH^-$) is believed to explain its long-term behavior and the decrease of its dissolution rate. The removal of open porosity in iodoapatites (either for lead-bearing or calcium-bearing) can be achieved for "open" systems for temperatures below 500 °C thanks to a liquid sintering, the efficiency of which is accentuated by SPS conditions. Vitrification can also be considered for the conditioning of iodine. However, this would require specific developments on the composition of silver phosphate glasses to improve their chemical durability. Such an objective could be achieved by the use of cross-linking reagents.

REFERENCES

1. *Control of iodine in the nuclear industry*, AIEA technical reports series No 148 (AIEA, 1973).
2. *Evaluation de la faisabilité du stockage géologique en formation argileuse, Dossier Argile 2005*, ANDRA (ANDRA, 2006), p. 211.
3. R.H. Nichols, C.M. Hohenberg, K. Kehm, Y. Kim and K. Marti K., *Geochemica et Cosmochimica Acta* **58**, 2553-2561 (1994).
4. C. Guy, F. Audubert, J.E. Lartigue, C. Latrille, T. Advorcat and C. Fillet, *C.R. Phys.* **3**, 827-837 (2002).
5. S. Le Gallet, L. Campayo, E. Courtois, S. Hoffmann, Yu. Grin, F. Bernard and F. Bart, *J. Nucl. Mat.* **400**, 251-256 (2010).
6. L. Campayo, A. Grandjean, A. Coulon, R. Delorme, D. Vantelon and D. Laurencin, *J. Mater. Chem.* **21**, 17609-17611 (2011).
7. H. Fujihara, T. Murase, T. Nisli, K. Noshita, T. Yoshida and M. Matsuda, *Mat. Res. Soc. Symp. Proc.* **556**, 375-382 (1999).
8. T. Sakuragi, T. Nishimura, Y. Nasu, H. Asano, K. Hoshino and K. Iino, Mater. Res. Soc. Symp. Proc. **1107**, 279-286 (2008).
9. T. Lemesle, F. Mear, L. Campayo, O. Pinet, B. Revel and L. Montagne, *J. Haz. Mat.* **264**, 117-126 (2014).
10. D. Grossin, S. Rollin-Martinet and C. Estournes, *Acta Biomaterialia* **6**, 577-585 (2010).
11. E.R. Maddrell and P.K. Abraitis, *Mat. Res. Soc. Symp. Proc.* **807**, 261-266 (2004).

Mater. Res. Soc. Symp. Proc. Vol. 1744 © 2015 Materials Research Society
DOI: 10.1557/opl.2015.393

Effects of hydrosulfide and pH on iodine release from an alumina matrix solid confining silver iodide

Tomofumi Sakuragi, Satoshi Yoshida, Osamu Kato[1], Kaoru Masuda[2]
Repository Engineering and EBS Technology Research Project, Radioactive Waste Management
Funding and Research Center, Tsukishima 1-15-7, Chuo City, Tokyo, Japan
[1]Kobe Steel, Ltd., Wakinohama-Kaigandori 2-2-4, Chuo-ku, Kobe, Japan
[2]Kobelco Research Institute, Inc., Takatsukadai 1-5-5, Nishi-ku, Kobe, Japan

ABSTRACT

Alumina matrix solidification is a hot isostatic pressing (HIP) technique used to immobilize radioactive iodine (^{129}I) in the form of silver iodide. In the present study, an alumina matrix solidification sample with a porosity of 12.9% was obtained by performing HIP at 175 MPa and 1200°C for 3 hours on a simulated spent silver-sorbent saturated with stable iodine. Material Characterization Centre-1 (MCC-1) leaching tests for the simulated waste form were performed using hydrosulfide (HS$^-$) as a reductant at concentrations ranging from 3×10^{-7} M to 3×10^{-3} M and at pH values ranging from 8.0 to 12.5. Leached iodine concentrations were below the detection limit for ICP-MS measurements at HS$^-$ concentrations of 3×10^{-7} M and 3×10^{-5} M. This result was due to the stability of AgI. At an HS$^-$ concentration of 3×10^{-3} M, iodine leaching rapidly increased within 10 days. The maximum iodine concentration in the solution was 4.33×10^{-3} M, which corresponds to 85% dissolution of the initial iodine. This value was measured after 552 days under an HS$^-$ concentration of 3×10^{-3} M at pH 11. An analysis of specimen cross-sections suggested the following reaction: $2AgI + HS^- = Ag_2S + 2I^- + H^+$. The pH affected matrix aluminum dissolution but did not significantly affect the iodine leaching behavior. Furthermore, the normalized mass loss of iodine was larger than that of aluminum by a factor greater than 10^4, which is due to the large porosity and the dissolution of interior AgI of the solid.

INTRODUCTION

The Japanese commercial reprocessing plant in Rokkasho is equipped with a unique and advanced system that prevents the release of volatile iodine into the outside environment. The system adsorbs iodine in the form of silver iodide (AgI) on an alumina-based filter on which silver nitrate has been deposited (the silver-sorbent). During 40 years of operation for 32,000 MTU (burn-up of 45 GWd/MTU), the Rokkasho reprocessing plant is estimated to have a total inventory of 5.1×10^{13} Bq of iodine-129 (^{129}I) and 150 tons of spent silver-sorbent [1]. The spent silver-sorbent is expected to be disposed of in a deep underground repository.

Ground-level handling and temporary storage of the spent silver-sorbent is convenient because AgI is one of the most stable iodides under oxidizing conditions. However, thermodynamic calculations have predicted that AgI will decompose into silver metal and iodide under reducing conditions [2]. In the performance assessment for radioactive waste disposal, which was described in a Japanese national report (the TRU-2 report) [1], the ^{129}I contained in spent silver-sorbent waste is assumed to be released instantaneously after disposal at depths greater than 300 m. This presumed release, combined with iodide migration, can strongly affect radiation doses. Therefore, an advanced waste form is expected to be developed in conjunction with the advanced reprocessing program to immobilize iodine. Two important requirements for this waste form are the efficient incorporation of iodine and long-term durability under repository conditions. In addition, a sound understanding of the iodine leaching behavior is necessary.

Several techniques have been used to immobilize iodine [3–11]. Potential materials for iodine immobilization in studies prior to 1987 were summarized by the IAEA [3]. Since then, the syntheses of certain iodine-bearing materials such as apatite [4, 5], Synroc [6], and zeolite have been demonstrated [7]. These works have focused on elucidating the behavior of iodine uptake into the matrix. Recent studies have involved durability and iodine leaching tests for waste-form stimulants under disposal conditions [8–11].

One potential method for physical confinement is to use alumina matrix solidification to retain fine AgI in the crystal grain boundaries of the α-alumina (corundum) matrix by subjecting the spent silver-sorbent to hot isostatic pressing (HIP) [11, 12]. This potential technique has two advantages: It is a simple process that does not isolate the adsorbed iodine, and the base material of the sorbent (alumina) can be used as a solidification matrix (corundum). The corundum matrix is expected to serve a protective role in AgI dissolution. One proposed solidification process consists of the following three steps: 1) pre-heating the spent sorbent to remove volatile components such as NO_x and O_2, 2) encapsulating the sorbent in a stainless steel vessel (25 L to 40 L) and vacuum sealing the vessel, and 3) performing HIP at 1200°C and 175 MPa for 3 hours [12]. This process has been demonstrated to result in a derived solid within a dense and robust body with a uniaxial compressive strength of 1.1 GPa [12].

The dissolution behavior of AgI [2, 13, 14] and the alumina matrix [11] have been investigated as fundamental processes to understand how they affect iodine leaching from the solid. The first set of studies showed that, among conceivable reductants, hydrosulfide ion (HS^-) strongly affected the dissolution of AgI. However, the most recent study demonstrated the pH dependence of alumina dissolution rates and provided an estimate of solubility changes when pH and cation species are changed. In this paper, we evaluate the effectiveness of alumina matrix solidification at preventing iodine leaching. Leaching tests were conducted across a range of HS^- concentrations and pH values, and subsequent tests were performed to characterize the solid surface after leaching. On the basis of these results, leaching processes are discussed.

EXPERIMENTAL

The immersion sample solidifications were prepared through the HIP of a simulated spent silver-sorbent. First, the simulated spent silver-sorbent was prepared by loading non-radioactive iodine onto a virgin silver-sorbent (provided from an industry) by passing through I_2 gas at 150°C in a column. Under this procedure, the amount of iodine adsorbed onto the silver deposits reached saturation. Notably, 15.3% of the silver nitrate loaded onto the virgin sorbent was altered into silver iodate and silver iodide (Table I). Next, the simulated spent sorbent was heated at 480°C for 3 hours to remove volatile components, crushed into 40 μm particles and packed into cylindrical stainless steel capsules (100 mL). The capsules, which had a vacuum nozzle attached, were evacuated with a rotary pump and sealed by welding. The HIP treatment was then performed at 1200°C and 175 MPa for 3 hours using an Ar-gas-pressurized HIP system equipped with a molybdenum furnace. The compositions of the virgin sorbent, simulated spent sorbent and solids produced by HIP were determined by XRD (Rigaku RAD-RU300), ICP-AES (Shimadzu ICPS-8000), and ion chromatography (DIONEX DX-500) [12].

The compositions of the materials are summarized in Table I. The resultant solid simply consists of α-Al_2O_3 (corundum) and AgI. The reduction of total iodine, as shown in this table, might be attributable to the volatilization of silver iodate during heating and evacuation.

Table I. Composition of chemical compounds loaded onto the sorbents (wt%)

	AgNO$_3$	Total iodine	AgIO$_3$	AgI
Virgin sorbent	15.3	-	-	-
Simulated spent sorbent (before thermal treatment) [12]	0.88	13.7	8.0	19
Simulated spent sorbent (after thermal treatment) [12]	< 0.1	10.7	0.1	20
Solidification after HIPing [12]	< 0.1	10.2	< 0.02	19

Immersion samples (10 mm cube) were cut from the HIP solid using a microtome and subsequently subjected to cross-section polishing. The bulk density and apparent density of the cube specimen were measured in accordance with Japanese Industrial Standard R 2205 (Archimedes' principle) and by using mercury porosimetry (Shimadzu AutoPore IV 9520). The open porosity and total porosity were calculated to be approximately 5% and 12.9%, respectively (Table II). The porosity of a sample cross-section was also estimated to be 13.2% through digital processing of an SEM image to measure the pore and matrix area.

Table II. Porosity of the alumina solidification

	Bulk density (g/cm^3)	Apparent density (g/cm^3)	Open porosity (%)	Total porosity (%)
JIS R 2205	3.83 ± 0.03	4.01 ± 0.07	4.3 ± 2.4	12.9 ± 0.8
Porosimetry	3.83	4.05	5.4	12.9
Digital image processing	-	-	-	13.2

Testing for aqueous leaching induced by hydrosulfide ion (HS$^-$) concentration, pH, and cation species was performed for approximately 500 days according to a Material Characterization Centre-1 (MCC-1) static low-temperature test protocol [15]. The block sample (10 mm cube) was immersed in a 600-mL solution (S/V ratio of 0.01 cm^{-1}) and kept in a glove box purged with N_2 gas (oxygen < 1 ppm); the temperature was maintained between 22°C and 23°C. The HS$^-$ concentration was controlled using an $Na_2S\cdot9H_2O$. The pH and Eh values of the solution were monitored during the test period and were adjusted as necessary to maintain the pH within ±1 unit of its initial range through the addition of conditioning agents. The concentrations of leached elements were measured using ICP-MS (PerkinElmer ELAN DRC II) after dilution of the samples with 0.02% sodium thiosulfate for iodine and 1% nitric acid for aluminum.

After approximately 500 days, several cross-section samples were observed by SEM-EDX (Hitachi SU-70 equipped with an Oxford Instruments INCA Energy +) and analyzed by XRD (Rigaku RAD-RU300). Immersion tests of the other samples continued after all the solutions were exchanged. The test conditions for the first and second immersions are summarized in Table III.

Table III. Conditions for leaching tests

No.	First immersion		Second immersion	
	HS^- (mol/L)	pH (reagent)	HS^- (mol/L)	pH (reagent)
Run 1	3×10^{-3}	12.5 (Ca(OH)$_2$)	3×10^{-3}	12.5 (NaOH)
Run 2	3×10^{-5}	12.5 (Ca(OH)$_2$)	-	-
Run 3	3×10^{-7}	12.5 (Ca(OH)$_2$)	3×10^{-3}	12.5 (Ca(OH)$_2$)
Run 4	3×10^{-3}	8 (HCl)	-	-
Run 5	3×10^{-3}	11 (-)	3×10^{-3}	11 (-)
Run 6	3×10^{-3}	12.5 (NaOH)	-	-

RESULTS AND DISCUSSION

Figure 1 shows the effect of hydrosulfide (HS^-) concentration on the iodine leaching behavior and on the Eh, pH, and Al concentration in the Ca(OH)$_2$ solution. The reducing conditions were controlled during the test period. The pH value increased slightly from 12.5 to a final value of 12.9. The aluminum concentration (Figure 1c) was close to the detection limit (3.7 $\times$ 10^{-7} M) until day 552, suggesting that Al dissolution is controlled by the solubility of certain calcium aluminates such as hydrogarnet (Ca$_3$Al$_2$(OH)$_{12}$). The thermodynamic calculator

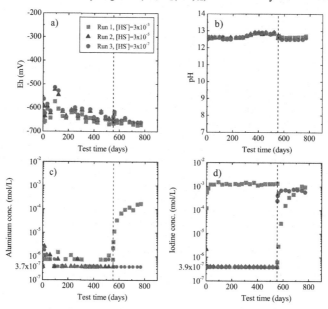

Figure 1. Effect of hydrosulfide concentration on a) Eh, b) pH, c) aluminum concentration, and d) iodine leaching behavior at pH 12.5. The dashed vertical line represents the time when the second immersed solution was exchanged (552 days). The values of 3.7 $\times$ 10^{-7} for Al and 3.9 $\times$ 10^{-7} for I represent the detection limits.

(PHREEQC ver. 2) predicted an Al concentration of 1.13×10^{-7} M at pH 13 (the final pH was approximately 12.8) in the Ca(OH)$_2$ solution. At an HS$^-$ concentration of 3×10^{-3} M (Run 1), the I concentration increased rapidly during the first 10 days and then remained approximately 1×10^{-3} M. In contrast, at HS$^-$ concentrations below 3×10^{-5} M (Runs 2 and 3), the concentration of iodine was below the detection limit. However, after the solution was changed from a 3×10^{-7} M solution to a 3×10^{-3} M solution (second dissolution test in Run 3), the I concentration increased rapidly and was similar to that observed in Run 1. Similarly, Inagaki *et al.* performed simple AgI dissolution tests and observed that AgI dissolution is strongly affected by the HS$^-$ concentration [13].

Figure 2 shows the effect of pH on the iodine leaching behavior during a test in which the Eh and pH of the solution were properly controlled. The dissolution of aluminum increased with increasing pH, which conforms to the thermodynamic calculation of Al solubility as a function of pH [11]. However, Al precipitation and secondary formed minerals have not yet been observed directly through experimentation. Iodine leaching was slightly correlated with Al dissolution.

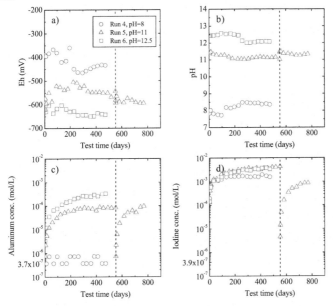

Figure 2. Effect of pH on iodine leaching behavior at a hydrosulfide concentration of 3×10^{-3} M.

Figure 3 shows cross-sections of the immersed samples. The gray area of the surface is an altered layer in which iodine is not present, whereas the inside patchy area is a pristine matrix. The thicknesses of the altered layer for Run 4 (475 days) and Run 5 (769 days) are approximately 0.5 mm and 2.5 mm, respectively. The corresponding equivalent altered layer thicknesses, T_{eq}, obtained from the leached iodine are 574 µm and over 5000 µm, respectively

(Table IV). The discrepancy in thickness measurements for Run 5 could be a consequence of the somewhat heterogeneous distribution of iodine.

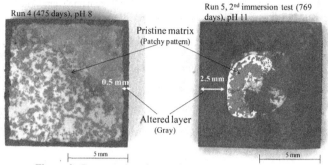

Figure 3. Cross-section observations of immersed samples.

Figure 4 shows an SEM micrograph of the boundary between the altered layer and the pristine matrix shown in Figure 3. The EDX line analyses of the elemental distributions from the altered to the pristine area are also shown. No contrast in appearance is evident between the altered and pristine area by SEM. However, the elemental distributions of sulfur and iodine clearly differ. In the altered area, the sulfur is collocated with silver; in contrast, iodine collocates with silver in the pristine area. Thus, the region where iodine is replaced by sulfur is identified as altered. This observation suggests that AgI dissolution and the subsequent iodine leaching are

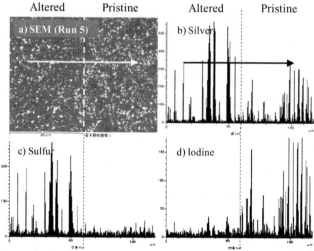

Figure 4. a) SEM observations near the boundary between the altered layer and pristine matrix; b), c) and d) are element distributions along the path indicated by the white arrow crossing the boundary in image a) (obtained by EDX line analysis).

attributable to the following reaction: $2AgI + HS^- = Ag_2S + 2I^- + H^+$. Inagaki *et al.* used XRD to confirm the formation of Ag_2S on the surface of AgI particles in contact with an Na_2S solution [13]. Moreover, the porosity was observed to increase from 13.2% in the pristine area to 20.8% in the altered area, which corresponds to a 53% reduction in volume from AgI (density: 5.67 g/cm^3) to Ag_2S (density: 7.33 g/cm^3). These results indicate that HS^- ions permeate the alumina matrix not only via the initial pores but also through the extended pores resulting from Ag_2S formation.

We obtained the normalized elemental mass losses, NL_i (g/m^2), from the following equation:

$$NL_i = \frac{C_i}{f_i} \cdot \frac{V}{S},$$

where C_i is the mass concentration of element i in the leachate (g/m^3), f_i is the mass fraction of element i in the specimen, V is the leachate volume (m^3), and S is the geometrical surface area of the specimen (m^2). As summarized in Table IV, NL_I is larger than NL_{Al} by a factor greater than 10^4 in cases where the dissolution of the alumina matrices is low. This incongruence between the leaching behavior of iodine and aluminum was caused by the dissolution of the internal AgI. As previously mentioned, the initial and extended pores could be a transport path for HS^- ions and dissolved iodine. The initial porosity of over 10% is slightly too large. One challenge for future research is to prepare a denser solid and confirm its long-term durability. Recent work has indicated that the porosity can be reduced nearly 5% by improving the pretreatment process [12]. The initial AgI content of 19% in the alumina matrix waste form could also be optimized in order to minimize the extension of the penetration pathway.

Table IV. Results of leaching tests.

No.	Time (d)	pH	Eh (mV)	Al conc. (mol/L)	I conc. (mol/L)	T_{eq} (μm)	NL_I / NL_{Al}
Run 1	552	12.59	-662	7.41×10^{-7}	1.34×10^{-3}	480	3.6×10^4
Run 1	769	12.69	-672	1.70×10^{-4}	9.45×10^{-4}	893	2.6×10^2
Run 2	552	12.61	-648	3.70×10^{-7}	$< 3.9\times10^{-7}$	< 0.13	< 20.8
Run 3	552	12.63	-652	3.70×10^{-7}	$< 3.9\times10^{-7}$	< 0.13	< 20.8
Run 3	769	12.57	-662	$< 3.7\times10^{-7}$	5.80×10^{-4}	196	$> 1.5\times10^4$
Run 4	473	8.36	-435	3.70×10^{-7}	1.57×10^{-3}	574	8.4×10^4
Run 5	552	11.01	-544	8.52×10^{-5}	4.33×10^{-3}	1005	1.0×10^3
Run 5	769	11.34	-594	9.63×10^{-5}	8.66×10^{-4}	> 5000	5.7×10^2
Run 6	473	12.08	-643	3.33×10^{-4}	3.39×10^{-3}	1514	2.0×10^2

CONCLUSION

In this work, the iodine leaching behavior of an alumina matrix solid as a function of hydrosulfide concentration and pH was investigated. Iodine leaching was negligible when the hydrosulfide concentration was less than 3×10^{-5} M. Iodine was leached only when the hydrosulfide concentration exceeded 10^{-3} M as a result of the reaction $2AgI + HS^- = Ag_2S + 2I^- + H^+$. In conclusion, the formation of Ag_2S enlarges the pores. Under rigorous conditions, the normalized mass loss of iodine is larger than that of aluminum by a factor greater than 10^4. The initial and enlarged pores provide a pathway for the migration of HS^- ions and dissolved iodine.

Improving waste forms to lower the porosity of the material and to optimize the AgI ratio to the alumina matrix is a challenge for future research.

ACKNOWLEDGMENT

This research is a part of the "Research and development of processing and disposal technique for TRU waste (FY2013)" program funded by Agency for Natural Resources and Energy, Ministry of Economy, Trade and Industry of Japan.

REFERENCES

1. Federation of Electric Power Companies (FEPC) and Japan Atomic Energy Agency (JAEA), Second Progress Report on Research and Development for TRU Waste Disposal in Japan (2007).
2. Y. Inagaki, T. Imamura, K. Idemitsu, K. Arima, O. Kato, H. Asano and T. Nishimura, J. Nucl. Sci. Technol. **45**, 859 (2008).
3. IAEA Technical Reports Series No. 276 Treatment, Conditioning and Disposal of Iodine-129 (1987).
4. F. Audubert, J. Carpena, J. L. Lacout and F. Tetard, Solid State Ionics **95**, 113–119 (1997).
5. M. C. Stennett, I. J. Pinnock and N. C. Hyatt, Mater. Res. Soc. Symp. Proc. **1475**, 221-226 (2012).
6. E. R. Vance and J. S. Hartman, Mater. Res. Soc. Symp. Proc. **556**, 41-46 (1999).
7. N. C. Hyatt, J. A. Hriljac, A. Choudhry, L. Malpass, G. P. Sheppard and E. R. Maddrell, Mater. Res. Soc. Symp. Proc. **807**, 1-6 (2004).
8. T. Sakuragi, T. Nishimura, Y. Nasu, H. Asano, K. Hoshino and K. Iino, Mater. Res. Soc. Symp. Proc. 1107, 279-285 (2008).
9. A. Mukunoki, T. Chiba, T. Kikuchi, T. Sakuragi, H. Owada and T. Kogure, Mater. Res. Soc. Symp. Proc. **1518**, 15-20 (2013).
10. Y. Haruguchi, S. Higuchi, M. Obata, T. Sakuragi, R. Takahashi and H. Owada, Mater. Res. Soc. Symp. Proc. **1518**, 85-90 (2013).
11. H. Miyakawa, T. Sakuragi, H. Owada, O. Kato and K. Masuda, Mater. Res. Soc. Symp. Proc. **1518**, 79-84 (2013).
12. K. Masuda, O. Kato, Y. Tanaka, S. Nakajima, T. Sakuragi and S. Yoshida, Progress in Nuclear Energy Special issue for Scientific Basis of Nuclear Fuel Cycle II 2014 (to be published).
13. Y. Inagaki, T. Imamura, K. Idemitsu, K. Arima, O. Kato, H. Asano and T. Nishimura, Mater. Res. Soc. Symp. Proc. **985**, 431-436 (2007).
14. M. Tada, Y. Inagaki, et al., Proceedings of GLOBAL 2011, Paper No. 446834 (2011).
15. Pacific Northwest Laboratory, MCC Workshop on Leaching of Radioactive Waste Form, NL Rep.-3318 (1980).

Mater. Res. Soc. Symp. Proc. Vol. 1744 © 2015 Materials Research Society
DOI: 10.1557/opl.2015.336

Evaluation of Sorption Behavior of Iodide Ions on Calcium Silicate Hydrate and Hydrotalcite

Taiji Chida[1], Jun Furuya[1], Yuichi Niibori[1] and Hitoshi Mimura[1]
[1]Dept. of Quantum Science & Engineering, Graduate School of Engineering, Tohoku University, 6-6-01-2 Aramaki-aza-Aoba, Aoba-ku, Sendai 980-8579 JAPAN

ABSTRACT

The migration retardation of anionic radionuclides, notably I-129, in radioactive waste repositories is one of the most critical factors for improving the performance of engineered barriers. To gain more fundamental knowledge required to make such improvements, this study examined the sorption behavior of iodide ions on calcium silicate hydrate (CSH) and hydrotalcite (HT), which act as anion exchangers. CSH was synthesized using CaO and fumed silica, with Ca/Si molar ratios ranging from 0.4 to 1.6. The weight ratio of CSH to HT was 1.0. These solid samples were immersed for 14 days in a 30 mL sample of pure water or 0.6 M NaCl solution, each of which contained 0.5 mM iodide ions with a given liquid/solid weight ratio (10, 15, or 20). Raman spectroscopy studies indicated that the structures of CSH and HT were maintained during the hydration of the solid phase and the sorption of iodide ions. The distribution coefficients for the sorption of iodide ions on CSH and HT ranged from 6 to 13 L/kg for pure water and from 1 to 2 L/kg for NaCl solution. These retardation effects for iodide ions would contribute toward improving the performance of the repository system as most conventional safety assessments assume that iodide ions hardly sorb on engineered barriers such as cementitious materials.

INTRODUCTION

Calcium silicate hydrate (CSH) is the main component of cementitious materials that are used for the construction of radioactive waste repositories. To assess the migration of radionuclides, it is important to understand the interactions between CSH and radionuclides. In particular, I-129 increases the dose rate for the biosphere because of its long half-life and low adsorption on engineered barriers [1]. However, it has been reported that iodide ions are incorporated into CSH during the formation process despite the anion species [2]. In addition, hydrotalcite (HT) also has the ability to exchange anions. This study discusses the addition of HT to CSH as a strategy to immobilize further iodide ions and evaluates the usefulness of HT as an aggregate for concrete including CSH. In this study, the sorption behavior of iodide ions in the presence of both CSH and HT was examined. To date, several studies have examined the interaction between radionuclides and CSH using CSH samples that were dried after being synthesized in liquid [3]. However, the use of dried CSH may not appropriately simulate the actual conditions in radioactive waste repositories, because the real engineered barrier would be saturated with groundwater after backfilling. Therefore, this study used CSH and HT in experiments without drying after their preparation.

EXPERIMENTAL

Hydrotalcite pretreatment

Synthetic HT with the same chemical formula as natural HT ($Mg_6Al_2(OH)_{16}CO_3 \cdot 4H_2O$) was purchased from Wako Pure Chemical Industries. Most HT samples had likely already adsorbed some anionic species because of the strong anion exchange ability of HT. Parker et al. reported that among monovalent species, OH^- shows the most selective sorption on HT followed by F^-, Cl^-, NO_3^-, and I^- [4]. Miyata reported that among anions, CO_3^{2-} sorption on HT is the most selective [5]. In fact, the synthesized HT used in this study also contained significant quantities of anions (OH^- and CO_3^{2-}). To remove these anions, the synthesized HT samples were heat-treated at 773 K [5], and X-ray diffraction (XRD) analysis of the heat-treated HT samples confirmed the removal of such anionic species. After cooling, the heat-treated HT samples were quickly added to pure water or 0.5 M NaCl solution. The HT structure was completely restored by hydration for 7 days, and its reconstruction was verified by XRD. These heat-treated HT samples were used in the following sorption experiments.

Procedures of sorption experiments

The sorption experiments in this study followed basic procedures reported elsewhere [2, 6]. CSH was synthesized by CaO and fumed silica (AEROSIL 300, Japan AEROSIL Ltd.). The Ca/Si molar ratios of CSH were set to 0.4, 0.8, 1.2, and 1.6. In samples with both CSH and HT, the weight ratio of CSH to HT was set to 1.0. Two different immersion solutions were tested: pure water (simulated fresh groundwater) and 0.6 M NaCl solution (simulated saline groundwater). The total volume of the immersion solution was 30 mL, and the liquid-to-solid weight ratio (L/S ratio) was set to 10, 15, or 20. Iodide ions were added in the following two ways: (1) iodide ions were added with the immersion solution to the solid samples (referred to as "co-precipitation samples") and (2) iodide ions were added after the hydration of CSH and HT in the immersion solution for 7 days (referred to as "surface sorption samples"). The iodide ion concentration was 0.5 mM. The mixing procedure was conducted in a glove bag saturated with nitrogen gas in order to avoid contact with air. The sample tubes were sealed tightly and shaken at 120 strokes/min at 298 K.

Fourteen days after the addition of the iodide ions, the iodide ion concentrations were measured. The liquid and solid samples were separated by centrifuging for 10 min at 7500 rpm and filtered with a 0.20-μm membrane filter. The concentration of iodide ions was measured by inductively coupled plasma atomic emission spectroscopy (SPS7800, Seiko Instruments Inc.). Raman spectra were collected using a laser Raman spectrophotometer (JASCO, NRS-3300QSE) equipped with a YAG laser (532 nm) for excitation.

The discussion herein focuses on some of the Raman peaks for CSH and HT (Table 1). The structure of CSH is similar to that of tobermorite and consists of a hydrated Ca–O layer between silicate chains. The Q^n units in Table 1, where n is the number of bridging oxygen atoms, describe the polymerization degree of the SiO_4 tetrahedrons in these silicate chains. For example, when the Q^1 peak increases and/or the Q^2 peak decreases (the ratio of Q^1/Q^2 increases), the silicate chains in CSH depolymerize. The peaks corresponding to Al–O–Mg and Al–O–Al stretching were also measured to verify the structure retention of HT.

Table 1. Raman spectral peaks focused on this study [7-9].

	Raman shft [cm^{-1}]	Assignation
CSH	950–1010	Symmetrical stretching of Q^2 tetrahedra
	870–900	Symmetrical stretching of Q^1 tetrahedra
Hydrotalcite	558	Al–O–Mg stretching
	490	Al–O–Al stretching

RESULTS AND DISCUSSION

Properties of the mixed CSH and HT sample

Figure 1 shows the volume of the liquid phase after the sorption experiments. Owing to the hydration of CSH and HT, the volume of liquid phase dramatically decreased from the initial volume of 30 mL. The decrease in the volume of the liquid phase became larger as the amount of solid increased. The addition of NaCl hardly affected the decrease in the volume of liquid phase (i.e., the hydration of CSH and HT). Figure 2 shows the relationship between pH values and Ca/Si ratios. The pH values of the mixed CSH and HT samples were significantly lower than those of the CSH only sample identical to those for L/S = 15 and 20. The pH differences between the samples with and without HT may be due to the buffering effect of HT, like the intercalation of OH⁻ in the layer structure of HT. Figure 3 shows the Ca/Si ratio of CSH for co-precipitation samples estimated by the concentration of Ca and Si in the liquid phase. Even though a small amount of Ca ions leached into liquid phase, the Ca/Si ratio of the solid phase (CSH) hardly changed. Furthermore, NaCl did not affect the leaching of Ca and Si ions. The surface sorption samples showed the same tendency.

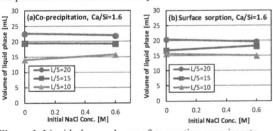

Figure 1. Liquid-phase volumes after sorption experiments.   **Figure 2.** pH of liquid phase.

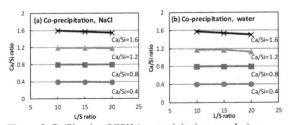

Figure 3. Ca/Si ratio of CSH (co-precipitation samples).

Sorption of iodide ions on CSH and HT

Figures 4 and 5 show the results of the sorption experiments. The amounts of iodine in Figures 4 and 5 do not indicate the iodide ion concentrations, but rather the amounts of substances (mol) depending on the volume changes of the liquid phase mentioned above. The initial amount of iodine was 15 μmol. The distribution coefficients K_d were estimated in the range of 6 to 13 L/kg for the pure water solution and in the range of 1 to 2 L/kg for NaCl solution. As shown in Figures 4 and 5, the sorption of iodide ions on the solid phase in NaCl solution was lower than that in pure water. In the authors' previous study, the values of K_d were estimated to be about 3 L/kg for the sorption of iodide ions on CSH in NaCl solution [2]; the values obtained under identical conditions in this study were almost the same. Therefore, the decrease in the sorption of iodide ions in NaCl solution is due to the effect of chloride ions on HT. As mentioned above, the order of the selective sorption of monovalent anions is $OH^- > F^- > Cl^- > NO_3^- > I^-$. Thus, in NaCl solution, the sorption of iodide ions on HT decreased owing to competition with chloride ions. In addition, the sorption amounts of iodide ions on surface sorption samples (Figure 5), which also depended on the Ca/Si ratio, exhibited more variation than those on co-precipitation samples (Figure 4). This suggests that iodide ions sorbed on solid samples are more unstable than those sorbed on the co-precipitation samples.

Raman spectroscopy analyses of the mixed CSH and HT samples

Figure 6 shows the Raman spectra of the solid phase mixed CSH and HT samples (L/S = 10 and Ca/Si = 1.6). The peaks emphasized by broken lines are described in Table 1. The spectra indicate Al–O–Al and Al–O–Mg stretching of HT along with the symmetrical stretching of the Q^1 and Q^2 tetrahedra of CSH. These results indicate that the structures of both HT and CSH were

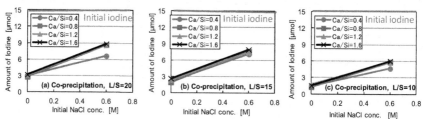

Figure 4. Sorption of iodide ions on CSH and HT (co-precipitation samples).

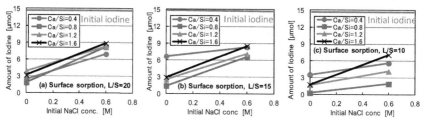

Figure 5. Sorption of iodide ions on CSH and HT (surface sorption samples).

maintained during the hydration of the solid phase and the sorption of iodide ions. Figures 7(a), (b), and (c) show the Q^1/Q^2 ratios estimated from the Raman spectra of the co-precipitation samples. The spectra of the surface sorption samples (not shown) showed the same tendencies. Figure 7(d) shows the Q^1/Q^2 ratios of CSH only (without iodide ions and HT, L/S = 20) [6]. The Q^1/Q^2 ratios of samples with low Ca/Si ratios (Ca/Si = 0.4 and 0.8) could not be estimated because the Q^1 peaks for these samples were not detected by Raman spectroscopy. As shown in Figures 7(a), (b), and (c), the silicate chain in CSH facilitated the polymerization of SiO$_4$ tetrahedrons along with the sorption of iodide ions in comparison with the non-iodide ion condition (Figure 7(d)). These results suggest that the sorption of iodide ions may contribute to the polymerization of silicate chains in CSH (i.e., the stabilization of CSH and the immobilization of iodide ions). On the other hand, NaCl did not have any influence on the Q^1/Q^2 ratios (Figure 7). Therefore, the presence of NaCl should not affect the structural changes in CSH more than the sorption of iodide ions does.

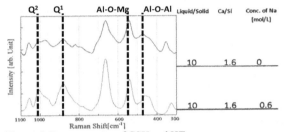

Figure 6. Raman spectra of CSH and HT.

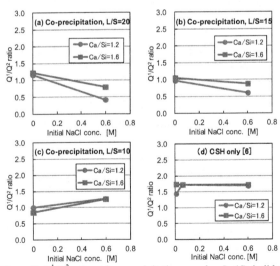

Figure 7. Q^1/Q^2 ratios of co-precipitation samples. In (d), iodide ions and HT were not added.

CONCLUSIONS

In this study, the sorption behavior of iodide ions on CSH and HT was discussed. In the experiments, the mixed CSH and HT samples efficiently sorbed iodide ions in an environment of pure water (the K_d values ranged from 6 to 23 L/kg). Although the sorption of iodide ions on HT decreased in NaCl solution owing to competition with chloride ions, the estimated K_d values for iodide sorption on CSH and HT ranged from 1 to 2 L/kg. These values are almost the same as those for CSH only. The results suggest that CSH and HT can act as efficient adsorbents for iodide ions in media ranging from pure water to NaCl solution. Raman spectra indicated that the structures of HT and CSH were maintained during the hydration of the solid phase and the sorption of iodide ions. In addition, the CSH structure would be stabilized by the sorption of iodide ions. These results indicate that CSH can immobilize iodide ions by means of sorption and stabilization. Moreover, the addition of HT as an aggregate for concrete including CSH will complement such retardation effects for anionic radionuclides, when CSH dissolves or hardly forms as a secondary mineral.

ACKNOWLEDGMENTS

This study was supported by the Japan Society for the Promotion of Science, Grant-in-Aid for Scientific Research (A) No. 25249136.

REFERENCES

[1] JNC (Japan Nuclear Cycle development institute), *H12 Project to Establish the Scientific and Technical Basis for HLW Disposal in Japan Vol. 3*, JNC TN1400 99 (1999).
[2] Furuya, J., Chida, T., Niibori, Y. and Mimura, H., *Proc. of WM 2014 Conference*, Paper No. 14077 (2014).
[3] Borrmann, T., Johnson, J.H., McFarlane, A.J., Richardson, M.J. and O'Connor, S.J., *J. Colloid Interface Sci.*, **339**, 175-182 (2009).
[4] Parker, L.M., Milestone, N.B., Newman, R.H., *Ind. Eng. Chem. Res.*, **34**, 1196-1202 (1995).
[5] Miyata, S., *Clays Clay Mineral.*, **31**, 305-311 (1983).
[6] Funabashi, T., Niibori, Y. and Mimura, H., *Proc. of WM 2012 Conference*, Paper No. 12145 (2012).
[7] Kirkpatrick, R.J., Yarger, J.L., McMillan, P.F., Yu, P. and Cong, X., *Adv. Cem. Based Mater.*, **5**, 93-99 (1997).
[8] Burrueco, M.I., Mora, M., Jimenez-Sanchidrian, C. and Ruiz, J.R., *J. Mol. Struct.*, **1034**, 38-42 (2013).
[9] Salam, M.A., Sufian, S. and Murugesan, T., *Mater. Chem. Phys.*, **142**, 213-219 (2013).

Mater. Res. Soc. Symp. Proc. Vol. 1744 © 2015 Materials Research Society
DOI: 10.1557/opl.2015.345

Study of the release of the fission gases (Xe and Kr) and the fission products (Cs and I) under anoxic conditions in bicarbonate water

Ernesto González-Robles[1], Markus Fuß[1], Elke Bohnert[1], Nikolaus Müller[1], Michel Herm[1], Volker Metz[1], Bernhard Kienzler[1]

[1] Karlsruhe Institute of Technology (KIT), Institute for Nuclear Waste Disposal (INE), Hermann-von-Helmholtz Platz 1, D-76344 Eggenstein-Leopoldshafen, Germany

ABSTRACT

For safety assessment analyses of the disposal of spent nuclear fuel (SNF) in deep geological repositories it is indispensable to evaluate the contribution of fission products to the instant release fraction (IRF). During the last three years the EURATOM FP7 Collaborative Project, "Fast / Instant Release of Safety Relevant Radionuclides from Spent Nuclear Fuel (CP FIRST-Nuclides)" was carried out to get a better understanding of the IRF.

Within CP FIRST-Nuclides, a leaching experiment with a cladded SNF pellet was performed in bicarbonate water (19 mM NaCl + 1 mM NaHCO$_3$) under Ar /H$_2$ atmosphere over 333 days. The cladded SNF pellet was obtained from a fuel rod segment which was irradiated in the Gösgen pressurized water reactor; the average burn-up of the segment was 50.4 MWd/kg$_{UO2}$. In the multi-sampling experiment, gaseous and liquid samples were taken periodically. The moles of the fission gases Kr and Xe released in the gas phase and those of ^{129}I and ^{137}Cs released in solution were measured. Cumulative release fractions of $(1.6 \pm 0.2) \cdot 10^{-1}$ fission gases, $(1.6 \pm 0.1) \cdot 10^{-1}$ ^{129}I and $(3.9 \pm 0.2) \cdot 10^{-2}$ ^{137}Cs, respectively, were achieved after 333 days of leaching. Accordingly the release ratio of fission gases to ^{129}I was 1:1 and the release ratio of fission gases to ^{137}Cs was 4:1, respectively.

INTRODUCTION

The disposal in deep bedrock repositories is considered as the preferred option for the management of spent nuclear fuel (SNF) in many countries [1-4]. The aim is to safely dispose of the highly radioactive material so that it is isolated from the biosphere for an appropriate length of time. A multi-barrier system is interposed between the SNF and the environment considering the SNF itself as the first technical barrier. Though, geological or geo-technical barriers may prevent to some extent groundwater contacting the fuel, intrusion of solutions into disposal rooms has to be taken into account within the long-term safety case of a SNF repository. Assessing the performance of SNF in a potential geological disposal system requires the understanding and quantification of the radionuclide release in case of water access. The alteration of SNF and the consecutive release of radionuclides involve the combination of many different processes, which can be grouped into two stages [5,6]: i) short term release of the so-called instant release fraction (IRF); ii) long term release controlled by the dissolution of the UO$_2$ grains, which is referred as matrix contribution.

Within the EURATOM FP7 Collaborative Project, "Fast / Instant Release of Safety Relevant Radionuclides from Spent Nuclear Fuel (CP FIRST-Nuclides)" the objective is to have a better comprehension of the IRF. The IRF is due to the segregation of a part of the radionuclide inventory to the gap interface between the cladding and the pellet, the pellet fractures as well as to fuel grain boundaries. Besides the fission gases Kr and Xe, volatile radionuclides such as ^{129}I,

^{137}Cs, ^{135}Cs, ^{36}Cl and ^{79}Se as well as radionuclides with other chemical properties such as ^{99}Tc, ^{107}Pd and ^{126}Sn are segregated [5]. The degree of segregation of the various radionuclides is highly dependent on in-reactor fuel operating parameters such as linear power rating, fuel temperature, burn-up and ramping processes. In the case of the fission gases, the gas release occurs by diffusion to grain boundaries, grain growth accompanied by grain boundary sweeping, gas bubble interlinkage and intersection of gas bubbles by cracks in the fuel [1]. The present paper focuses on the release behavior of ^{137}Cs, ^{129}I and of fission gases Xe and Kr from spent nuclear fuel in contact with near neutral pH bicarbonate water under hydrogen overpressure.

EXPERIMENTAL

Spent nuclear fuel

The studied SNF was irradiated in four cycles during 1226 days at the Gösgen nuclear power plant located in Switzerland. The fuel achieved an average burn-up of 50.4 MWd/kg$_{UO2}$ and the average linear power was 260 W·cm^{-1} during irradiation. The cooling time of the fuel prior to start of the experiment was 24 years. Characteristic data of the studied fuel rod segment are given in [7, 8]. The fission gas release from the plenum of the fuel rod segment was previously measured to be (8.35 ± 0.66) % [8,9]. A cladded fuel pellet (figure 1) was cut from the fuel rod in a hot cell under N$_2$ atmosphere with an O$_2$ content < 1% [8,10]. The cutting was performed in the gap between two adjacent pellets taking into account the γ-scan of the fuel rod as reference. The complete process is described elsewhere [10]. Prior to start of the experiment, the sample was cut and the geometric properties of the cladded SNF pellet were characterized. The measurements of the external and internal diameters of the sample were (10.75 ± 0.01) mm and (9.35 ± 0.01) mm, respectively. The length was (9.85 ± 0.01) mm and the weight was (7.77 ± 0.01) g. Finally, taking into account the density of the fuel (10.41 g·cm^{-3}), length, internal diameter and mass of the fuel (without cladding) was derived to be (6.97 ± 0.06) g.

Figure 1. Image of the studied SNF pellet with Zircaloy cladding.

Experiment

A static leaching experiment with the cladded pellet was carried out in a 250 mL stainless Ti-lined VA steel autoclave (Berghof Company, Eningen, Germany) with two valves in the lid to allow the sampling of gas and solution. The leachant was prepared in a glove-box under Ar atmosphere with a final composition of 19 mM NaCl + 1 mM NaHCO$_3$, a pH$_0$ of (8.9 ± 0.2) and Eh$_0$ of (-116.3 ± 50) mV. The experiment was carried out under Ar/H$_2$ atmosphere with a hydrogen partial pressure of (3.0 ± 0.1) bar $p(H_2)$ and an argon partial pressure of (37.0 ± 1) bar $p(Ar)$. The sample was mounted in a Ti sample holder to ensure the contact of both pellet surfaces with the solution. Once the sample was placed into the autoclave, the autoclave was closed and purged with Ar to remove completely air out of the reaction vessel.

A pre-leaching test with 1 day duration was performed by filling the autoclave with 220 mL of bicarbonate water under Ar flux. This pre-leaching test was performed to dissolve any pre-oxidation layers potentially present at the sample surface as well as to reduce the amount of ^{135}Cs and ^{137}Cs in solution. The reduction of the ^{135}Cs and ^{137}Cs concentrations and thereby the reduction in the γ-dose rate was required to analyze solution aliquots without extensive dilution and to use the aliquots without remote handling. From the pre-leaching test, a solution aliquot and a gas sample were collected and analyzed.

After pre-leaching, the solution was completely replenished. Then, the autoclave was again refilled with 220 mL of fresh bicarbonate water. During the consecutive static leaching experiment, gaseous $(50 \pm 1$ mL$)$ and liquid $(15 \pm 1$ mL$)$ samples were taken after 1, 7, 21, 56, 84, 176, 245 and 332 days. After the sampling, the gas volume of the autoclaves was purged with Ar, and the initial conditions (40 bar of Ar + H_2 mixture) were again established. The solution was not renewed after sampling. Therefore, the remaining leachant volume was reduced at each sampling step.

Analyses of released radionuclides in liquid and gaseous samples

The collection of the gas samples was done by using stainless steel single-ended miniature sampling cylinders (SS-4CS-TW-50, Swagelok Company, USA). These samples were used to determine the amount of Kr and Xe released during the leaching experiments as well as to monitor the gas atmosphere (e.g. checking for potential air intrusion). Using the volume of the cylinders (50 mL) and the results of the gas-mass-spectrometry, the moles of released gases were obtained.

From the liquid samples obtained during each sampling campaign different aliquots were prepared to determine the amount of ^{129}I and ^{137}Cs:

- An aliquot of 1 mL was prepared by diluting and acidifying 0.1 mL of sample with 0.9 mL of 1M HNO_3 solution. The activity of ^{137}Cs in the aliquot was measured by γ-spectrometry by means of Ge-detectors (EGC-15-185-R and GX3018, Canberra Industries Inc, Meriden, USA).
- Another aliquot of 1 mL was prepared by precipitating the amount of Cs present in 2 mL of sample with ammonium molybdophosphate (AMP). It is emphasized that the pH was constant during the AMP treatment, because the solution was not acidified. By means of the AMP precipitation methods the activity of ^{137}Cs in the sample was considerably reduced and thereby the determination of ^{129}I by γ-spectrometry was improved [11].

The fraction of the inventory released either in the gas or the aqueous phases was derived taking into account the inventory of the samples. Based on the characteristic fuel and irradiation data given in [7], inventories of ^{137}Cs, ^{129}I and of fission gases Xe and Kr were calculated using the *webKorigen* software package [12]: $5.3 \cdot 10^{-6}$ mol$\cdot(g_{UO2})^{-1}$ ^{137}Cs, $1.6 \cdot 10^{-6}$ mol$\cdot(g_{UO2})^{-1}$ ^{129}I and $5.8 \cdot 10^{-6}$ mol$\cdot(g_{UO2})^{-1}$ fission gases (sum of Kr and Xe radioisotopes), respectively. In case of the FG inventory of the fuel rod segment plenum already released during puncturing test would not be considered, the FG inventory is only $5.3 \cdot 10^{-6}$ mol$\cdot(g_{UO2})^{-1}$.

RESULTS AND DISCUSSION

The number of moles released in the gas phase of the fission gases (Kr + Xe) as well as those released in solution of [129]I and ^{137}Cs are shown in figure 2. It can be observed that there is a high release of [129]I and ^{137}Cs during the pre-leaching test (closed symbols in figure 2). This observation is explained by the fact that these two relatively volatile radionuclides tend to migrate during irradiation to void spaces of the SNF rod such as fractures, grain boundaries and the gap between the pellet and the cladding. Once the bicarbonate water reaches the fuel, the accessible fraction of the [129]I and ^{137}Cs inventories are fast/instantaneously released to solution. Since (8.35 ± 0.66) % of the FG inventory, which had been present in the plenum of the fuel rod segment, was already released in the puncturing test prior to the leaching, a high release of the fission gases in pre-leaching test was not observed.

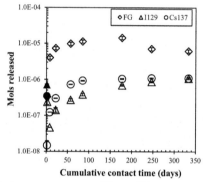

Figure 2. Moles released of fission gases (FG), [129]I and ^{137}Cs from the studied SNF as a function of time. Closed symbols represent moles released in the pre-leaching test.

After the pre-leaching test, the moles of dissolved [129]I and ^{137}Cs steadily increased during the first 85 days. Afterwards the amount of released [129]I slightly increases and ^{137}Cs virtually achieved a steady state until the end of the experiment. In contrast, an increase in the amount of fission gases released was observed during the initial phase of 21 days of leaching, followed by a steady state until 176 days. According to the last two FG samples, the amount of released fission gases was lower than the steady state values. It has to be noticed that the gas phase is replaced after each sampling. These pseudo-dynamic conditions caused the reduction of the released FG moles at the end of the experiment.

In order to evaluate how much of the inventory has been released through the experiment, the fraction of inventory released for an element i was calculated following equation (1):

$$Fraction\ release = \frac{m_i}{m_{UO_2} x H_i} \tag{1}$$

where m_i is the mass of element i (g) in the gas or liquid phase, m_{UO2} is the initial oxide mass (g) in the fuel sample and H_i corresponds to the fraction of inventory for the element i ($g_i \cdot g_{UO2}^{-1}$). The elemental inventories of the studied SNF were calculated using the *webKorigen* software package [12]. In the case of the FG, the considered inventory was the initial inventory ($5.8 \cdot 10^{-6}$ mol·g_{UO2}^{-1}) before the release of the FG plenum inventory in the puncturing test.

Taking into account the fractional release according to equation (1), the cumulative fraction of released inventory was calculated as the summary of the release for each contact period as described in equation (2):

$$Cumulative\ release\ fraction = \sum Fraction\ release_i \qquad (2)$$

In figure 3, the cumulative release fraction of FG, [129]I and ^{137}Cs as function of the cumulative contact time is plotted. The cumulative fraction released from the inventory of the cladded fuel pellet after 333 days of leaching time was $(1.6 \pm 0.2) \cdot 10^{-1}$ for FG, $(1.6 \pm 0.1) \cdot 10^{-1}$ for [129]I and $(3.9 \pm 0.2) \cdot 10^{-2}$ for [137]Cs. As mentioned earlier, the release fraction is slightly higher for FG, [129]I and ^{137}Cs during the first 85 days of leaching. The release of [129]I is higher than the ^{137}Cs release as it was previously observed in former studies [5,11,13]. Based on the cumulative release fraction, the release of fission gases and [129]I is virtually equal after 56 days leaching time. Contrarily, the release ratio of fission gases to ^{137}Cs is about 4:1.

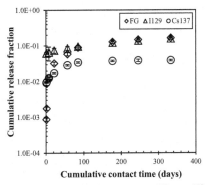

Figure 3. Cumulative release fraction of fission gases (FG), [129]I and ^{137}Cs from the studied SNF as a function of leaching time.

Finally, the fractional release rate (d^{-1}) was calculated taking into account the amount of radionuclide released during each contact period of time. In figure 4, the fraction release rate of FG, [129]I and ^{137}Cs as a function of the cumulative contact time is shown.

The fractional release rate is steadily decreasing for FG, [129]I and [137]Cs. After 87 days, a relatively strong decrease in the ^{137}Cs fractional release rate is observed. According to our interpretation the fast leachable ^{137}Cs fraction is substantially decreased with time to such an extent that the release of ^{137}Cs would be terminated. In the case of FG and [129]I, the fractional release rate at the end of the experiment is significantly slower than at the beginning, approaching virtually to a steady state. Still, FG and [129]I are released at a fractional release rate of about $2 \cdot 10^{-4}$.

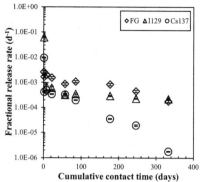

Figure 4. Fractional release rate of fission gases (FG), ^{129}I and ^{137}Cs from the studied SNF as a function of the cumulative contact time.

CONCLUSIONS

The release of ^{129}I, ^{137}Cs and of fission gases from a SNF sample in near neutral pH bicarbonate water was studied under 40 bars of Ar/H$_2$ atmosphere (pH$_2$: 3 bar). The multi-sampling experiment was performed with a cladded SNF pellet with an average burn-up of 50.4 MWd/kg$_{UO2}$. Analyses of released ^{129}I, ^{137}Cs and of fission gases demonstrate:

- A high release of ^{129}I and ^{137}Cs during a 1 day pre-leaching test.
- A high release of FG, ^{129}I and ^{137}Cs during the first 87 days of the static leaching experiment. The cumulative release after 333 days of leaching was: $(1.6 \pm 0.2) \cdot 10^{-1}$ for FG, $(1.6 \pm 0.1) \cdot 10^{-1}$ for ^{129}I and $(3.9 \pm 0.2) \cdot 10^{-2}$ for ^{137}Cs
- The ratio between the FG and ^{129}I and ^{137}Cs was 1:1 and 4:1, respectively.
- The fractional release rate indicates a considerably low release of ^{137}Cs after 87 days of leaching. Virtually a steady state release of FG and ^{129}I with a fractional release rate of about $2 \cdot 10^{-4}$ was observed after 245 days.

Acknowledgement

The research leading to these results has received funding from the European Union's European Atomic Energy Community's (Euratom) Seventh Framework Programme FP7/2007-2011 under grant agreement n° 295722 (FIRST-Nuclides project).

References

1. L. Johnson and D.W. Shoesmith. "Spent Fuel", *Radioactive Waste Forms for the Future*, ed. W. Lutze and R.C. Ewing (North-Holland, 1988) pp. 635-698.
2. D.W. Shoesmith, J. Nucl. Mater. 282, 1-31(2000).
3. J. Bruno and R.C. Ewing, Elements 2, 343-349 (2006).

4. V. Metz, H. Geckeis, E. González-Robles, A. Loida, C. Bube and B.Kienzler, Radiochim. Acta **100** (8-9), 699-713(2012).
5. L. Johnson, C. Ferry, C. Poinssot and P. Lovera, J. Nucl. Mater. 346, 56-65(2005).
6. C. Poinssot, C. Ferry, P. Lovera, C. Jegou and J.-M. Gras J. Nucl. Mater. 346, 66-77 (2005).
7. V. Metz, E. González-Robles, and B. Kienzler, in *2nd Annual Workshop Proceedings of the Collaborative Project "Fast/Instant Release of Safety Relevant Radionuclides from Spent Nuclear Fuel"*, edited by. B. Kienzler, V. Metz, L. Duro and A. Valls (Karlsruhe KIT-SR 7676, 2014) pp. 55-60.
8. E. González-Robles, D. H. Wegen, E. Bohnert, D. Papaioannou, N. Müller, R. Nasyrow, B. Kienzler, V. Metz, in *Scientific Basis for Nuclear Waste Management XXXVII*, edited by L. Duro, J. Giménez, I. Casas, J. de Pablo ((Mater. Res. Soc. Symp. Proc. **1665**, Barcelona, Spain, 2014) pp.283-289.
9. E. González-Robles, E. Bohnert, N. Müller, M. Herm, V. Metz and B. Kienzler, in *2nd Annual Workshop Proceedings of the Collaborative Project "Fast/Instant Release of Safety Relevant Radionuclides from Spent Nuclear Fuel"*, edited by. B. Kienzler, V. Metz, L. Duro and A. Valls (Karlsruhe KIT-SR 7676, 2014) pp. 37-40.
10. D.H. Wegen, D. Papaioannou, R. Gretter, R. Nasyrow, V.V. Rondinella, J.-P. Glatz, in *1st Annual Workshop Proceedings of the Collaborative Project "Fast/Instant Release of Safety Relevant Radionuclides from Spent Nuclear Fuel"* edited by. B. Kienzler, V. Metz, L. Duro and A. Valls (Karlsruhe KIT-SR 7639, 2013) pp. 193-199.
11. L. Johnson, I. Günther-Leopold, J. Kobler Waldis, H.P. Linder, J. Low, D. Cui, E. Ekeroth, K. Spahiu and L.Z. Evins, J. Nucl. Mater. 420, 54–62 (2012).
12. Nucleonica GmbH (2011) *Nucleonica Nuclear Science Portal (www.nucleonica.com)*, Version 3.0.11.
13. O. Roth, J. Low, M. Granfors, K. Spahiu, in *Scientific Basis for Nuclear Waste Management XXXVI*, edited by N. C. Hyatt, K. M. Fox, K. Idemitsu, C. Poinssot, K. R. Whittle ((Mater. Res. Soc. Symp. Proc. **1518**, Boston, MA, 2013) pp. 145-150.

Mater. Res. Soc. Symp. Proc. Vol. 1744 © 2015 Materials Research Society
DOI: 10.1557/opl.2015.310

Technetium Getters to Improve Cast Stone Performance

James J. Neeway, Amanda R. Lawter, R. Jeffrey Serne, R. Matthew Asmussen and Nikolla P. Qafoku

Pacific Northwest National Laboratory, Energy and Environment Directorate

Richland, WA 99352, U.S.A.

ABSTRACT

A cementitious waste form known as Cast Stone is the baseline waste form for solidification of aqueous secondary wastes, including Hanford Tank Waste Treatment and Immobilization Plant (WTP) secondary liquid effluents to be treated and solidified at the Hanford Site Effluent Treatment Facility. Cast Stone is also being evaluated as a possible supplemental immobilization technology to provide the necessary low activity waste (LAW) treatment capacity to complete the Hanford tank waste cleanup mission in a timely and cost effective manner. Two radionuclides of particular concern in these waste streams are technetium-99 (^{99}Tc) and iodine-129 (^{129}I). These radioactive tank waste components are predicted to contribute the most risk to groundwater – the most probable pathway for future environmental impacts associated with the cleanup of the Hanford site. A recent environmental assessment of Cast Stone performance, which assumes a diffusion controlled release of contaminants from the waste form, calculates groundwater in excess of the allowable maximum permissible concentrations for both contaminants. There is, therefore, a need and an opportunity to improve the retention of both ^{99}Tc and ^{129}I in Cast Stone. One method to improve the performance of Cast Stone is through the addition of "getters" that selectively sequester Tc and I, therefore reducing their diffusion out of Cast Stone. In this paper, we present results of Tc and I removal from solution with various getters. Batch sorption experiments were conducted with deionized water (DIW) and a highly caustic LAW simulant with a 7.8 M average Na concentration. In general, the data show that the selected getters are effective in DIW but their performance is compromised when experiments are performed with the 7.8 M Na Ave LAW simulant. The diminished performance in the LAW simulant may be due to competition with Cr present in the 7.8 M Na Ave LAW simulant and to a pH effect that may create a negatively charged surface that can repel negatively charged species.

INTRODUCTION

For the last several decades, Portland cement grouts have been studied as encapsulation materials for low-level radioactive waste because of flexibility in processing, good durability, and the ability to tailor the properties of the final product [1]. One such grout, which is the baseline waste form for solidification of aqueous secondary liquid effluent from the Hanford Waste Treatment and Immobilization Plant (WTP), is Cast Stone. The low-temperature Cast Stone waste form is also being evaluated as a possible supplemental immobilization technology to increase Hanford LAW treatment capacity to complete the tank waste cleanup mission in a timely and cost-effective manner. The Cast Stone baseline dry blend mix currently used consists of 8 wt% Portland cement Type I/II, 45% Class F fly ash, and 47% Grade 100 or 120 blast furnace slag (BFS). A secondary liquid waste or LAW feed is then mixed with the dry reagents at a given free water to dry blend solids ratio (usually 0.6) and gravity-fed into containers for curing [2]. A similar waste form called Saltstone is used at the Savannah River Site (SRS) to solidify its LAW tank wastes [1; 2]. The Saltstone formulation differs by only a few percent in the dry blend mix with 45 wt% Grade 100 slag cement, 45 wt% Class F fly ash, and 10 wt% Type I/II Portland cement, which has been used since the late 1980s with little change to the formulation. The cementitious waste forms consist primarily of an amorphous hydrous calcium-silicate gel (C-S-H) and a few crystalline phases. When compared to the homogeneous borosilicate glass expected to be produced at WTP, Cast Stone has the added complexity that it changes from the smallest scale (crystal structure) through the meso-scale (microstructure, porosity, surface area) to the macro-scale (density, strength). However, small differences at the smallest scale have been shown to give rise to no significant effects at the other scales that may impact the waste form performance [3].

At the Hanford Site, the current plan is to dispose the immobilized low-activity waste (ILAW) and the solidified secondary wastes (i.e. Cast Stone) in the Integrated Disposal Facility (IDF). Due to their high environmental mobility and intermediate half-lives, two radionuclides that may have the biggest impact on safe disposal of Cast Stone at the IDF are ^{99}Tc ($t_{1/2} = 2.1 \times 10^5$ a) and ^{129}I ($t_{1/2} = 1.57 \times 10^7$ a). Past performance assessments and risk assessments, which are used to evaluate the potential impact from disposal of various solidified LAW forms, have shown that releases from non-glass waste forms may not meet environmental protection standards over long time periods. These assessments for cementitious (grout/Cast Stone) waste forms assume that release of ^{99}Tc and ^{129}I from the cementitious waste form is diffusion controlled. The effective diffusivities, which are a combination of physical and chemical interactions in the material, used in the assessments ranged from 3×10^{-10} cm^2/s to 5×10^{-9} cm^2/s for Tc and 1.0×10^{-10} cm^2/s to 2.5×10^{-9} cm^2/s for I [4; 5]. Because Cast Stone disposal may not meet environmental standards with the given Tc and I diffusivities, a need may exist for improved retention of these radionuclides within the Cast Stone formations. One possibility for improved retention is with the use of getters, which are inorganic materials that selectively absorb radionuclides and metallic contaminants.

In the present paper we give an overview of getter effectiveness for Tc and I removal from solution using three Tc getter materials (two blast furnace slags and Tin(II) Apatite) and two I getters (argentite and silver zeolite). These getters have previously been identified as having the potential to effectively bind Tc or I and limit their release from Cast Stone [6]. However, the majority of those studies that have examined potential getter materials were performed in circumneutral pH conditions with relatively low ionic strength solutions compared to the secondary and supplemental waste streams that are expected to be produced as a result of Hanford cleanup activities. We note that one set of experiments has previously utilized Tin(II) Apatite as a getter in the Cast Stone formulation [7]; however, the exact mechanism and location of Tc in the resulting material remain unknown. The present set of experiments serves both to canvass candidate getter materials and to elucidate the mechanism(s) for removal. Additional results from similar studies can be found elsewhere [8-10].

EXPERIMENT

Three different Tc getters and two different I getters have been tested. Information about the different getter materials is presented in Table 1. Two of the three Tc getters are BFS; BFS 1 is from a Northwest USA source and BFS 2 is from a Southeast USA source and is currently being used for Saltstone fabrication. The reducing potential of both BFS has already been investigated [11]. The BFS 2 contains less Al_2O_3 (6.6–8.4% vs. 12.2%), less CaO (35–38.5% vs. 43.4%), and less SO_3 (0.3–2.08% vs. 4.9%), but more MgO (12.9–13.1% vs. 4.9%) than BFS 1 [12]. One of the iodine getters, argentite, was synthesized in the laboratory following an established procedure [13].

Table 1. List of the getter materials used in the experiment.

Targeted Element	Getter	Full Name or Chemical Formula	Vendor Name
Tc	BFS 1	Blast Furnace Slag NW	Lafarge North America
Tc	BFS 2	Blast Furnace Slag SE	Holcim (US) Inc.
Tc	Tin Apatite	$Sn_5(PO4)_3(F,Cl,OH)$	Outside Laboratory
I	Argentite	Ag_2S	Synthesized in lab
I	Ag Zeolite	Silver-exchanged zeolite	Aldrich

Testing involved placing 1.0 g of getter material in contact with 100 mL of solution for periods up to 71 days with periodic solution sampling. To begin the experiment, the solution was spiked with a solution of ~52 ppm Tc (as $NaTcO_4^-$) or ~6.5 ppm I (as NaI). The Tc and I tests were run separately (i.e., the solutions did not contain both solutes). The concentrations were chosen to be are10× more concentrated than the projected waste stream and, thus, this allows more sensitivity in measuring the resulting Tc and I concentrations. The small (2 mL) liquid phase volume removed during each sampling was not replaced. Care was taken to ensure that none of the solid material was removed from the vial during sample collection. A 1-mL aliquot of the aliquot, filtered with a 0.45 μm membrane filter, was taken and analyzed by inductively coupled plasma-mass spectroscopy (ICP-MS) for concentration measurements. The two different solution media were 18.2 MΩ DI H_2O (DIW) and the 7.8 M Na LAW simulant—the simulant selected to represent an average LAW in Hanford Site single-shell tanks. Each test was run in duplicate and conducted in a polytetrafluoroethylene (PTFE) bottle at room temperature (~22 °C) in an anoxic chamber containing N_2 with a small amount of H_2 (0.7%) to maintain anoxic conditions. Oxygen levels within the chamber, which spiked briefly after introducing materials into the chamber, were measured near 5 ppm throughout the test. At the end of the experiment, the solid getter material was separated from the solution, dried in the anoxic chamber and examined by Scanning Electron Microscopy/Energy Dispersive

Spectroscopy (SEM/EDS). The instrument used was a FEI Helios 600 NanoLab FIB-SEM. Operating conditions were typically 5 keV or less for imaging and 20 keV for EDS measurements. EDS spectra were collected using Oxford INCA software with a live count of 100 seconds with a typical dead time of 30%.

The 7.8 M Na Ave LAW simulant was developed based on Hanford Tank Waste Operation Simulator (HTWOS) model runs, which project the future feed vector to a supplemental immobilization facility, and the chemical composition is presented in Table 2. This is the simulant used in the LAW Cast Stone screening tests that was named 7.8 M Na Ave [2]. For 7.8 M Na Ave experiments performed without Cr, the same simulant production method was used, but the Cr salt addition step was omitted. It should be noted that a relatively small amount of undissolved solids remained at the bottom of the vessel when the solution preparation was complete. The solids were most likely Na-phosphate, Na-fluoride and small amounts of Ni salt.

Table 2. The 7.8 M Na Ave simulant used in the batch sorption experiments.

Waste Constituent	Overall Average (mol/L)	Waste Constituent	Overall Average (mol/L)	Waste Constituent	Overall Average (mol/L)
Al^{3+}	0.48	F^-	0.05	TOC total[*]	0.12
K^+	0.06	NO_2^-	0.88	Free OH^-	2.43
Na^+	7.8	NO_3^-	2.53	$Cd^†$	0.25
Cl^-	0.06	PO_4^{3-}	0.08	$Cr^†$	33.3
CO_3^{2-}	0.43	SO_4^{2-}	0.13	$Pb^†$	0.4

[*]Added as oxalate to the simulant
[†]Concentration is in mmol/L

RESULTS AND DISCUSSION

<u>**Batch sorption tests for Tc**</u>

The assumed mechanism for Tc removal from solution is Tc(VII) reduction and transformation from the dominant aqueous pertechnetate [$Tc(VII)O_4^-$] species to the less soluble Tc(IV) species. This reaction requires the transfer of electrons from the getter to the pertechnetate ion. However, the 7.8 M Na Ave simulant contains substantial amounts of aqueous Cr(VI) (~20,000 ppm) that may also be reduced from the higher oxidation state to Cr(III). This competitive electron transfer reaction may diminish the effectiveness of the getter material when the contacting solution contains Cr(VI). Another area of concern is the high pH of the 7.8 M Na Ave simulant (~13.4). Under this condition some getter materials may develop a negative surface charge that can repel the negatively charged pertechnetate ion and, therefore, inhibit its potential sorption and reduction reactions occurring either at or near the getter surface.

In this set of experiments, we obtained results investigating both the effect of competition with Cr and the high pH environment. Experiments of time-dependent Tc removal from DIW, the 7.8 M Na Ave simulant with Cr, and the 7.8 M Na Ave simulant without Cr were performed with BFS 1, BFS 2, and Tin Apatite and are presented in Figure 1. In the experiments conducted with DIW, Tc removal was best for Tin Apatite, followed by BFS 2, and then BFS 1. For results with Tin Apatite, over 99% of the original Tc in solution was removed in the first three days of contact. The removal of Tc by the BFS materials was slower. However, after 34 days of contact, BFS 2 removed over 99% percent of Tc from solution and BFS 1 removed 66% of the original Tc. For BFS 1 the trend seems to indicate that more Tc would be removed if the reaction were allowed to further continue.

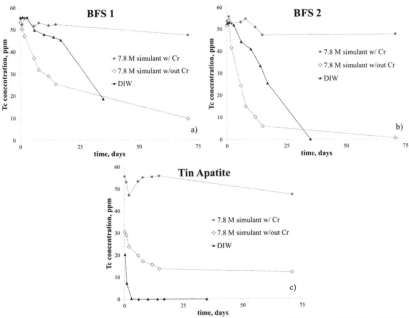

Figure 1. The removal of Tc from solution as a function of time for a) BFS 1, b) BFS 2, and c) Tin Apatite. Results are presented for experiments performed in DIW, 7.8 M Na Ave simulant with Cr, and 7.8 M Na Ave simulant without Cr. The initial Tc spike is roughly 52 ppm.

The promising results for Tc removal from DIW were not observed in the presence of the 7.8 M Na Ave simulant containing roughly 20,000 ppm Cr. For all of the Tc getters in this solution, less than 15% of the original Tc was removed from solution after 71 days of contact.

When the experiment was performed with the 7.8 M Na Ave simulant without Cr, after 71 days of contact, Tin Apatite removed 75 % of the original Tc, and BFS 1 and BFS 2 removed more than 99% of the original Tc. In fact, for the two BFS materials, the Tc removal was faster when compared with Tc removal from DIW. This enhanced removal in the highly caustic solution compared to DIW may be from an increase in the dissolution of solid phases present in the BFS materials that released reductants such as sulfide and ferrous iron into solution. In fact, the dissolution of BFS and the release of these reductants has been observed to increase the reduction of Tc(VII) to Tc(IV) upon curing of the Cast Stone waste form [11]. On the other hand, for Tin Apatite the removal of Tc from the 7.8 M Na Ave solution without Cr was less than the removal of Tc from DIW. The decreased effectiveness of Tin Apatite may merely be a result of the high pH [7], however, it is not possible to exclude the possibility that most of the surface Sn(II) was oxidized to Sn(IV) and further reduction of remaining aqueous Tc(VII) was not possible. This observation is supported by minimal change in the Tc concentration after 15 days of contact indicating that the Tc(VII)-reducing reaction was at equilibrium. However, it would be necessary to further examine the post-reaction Tin Apatite solid in contact with the Tc solution to verify this observation.

Batch sorption tests for I

For iodide removal from solution, the best-performing getter materials are generally silver-based and rely on the formation of the largely insoluble silver iodide (AgI). At this point in these studies, we have not investigated if either of the BFS materials are capable of removing I from solution. The batch sorption experiments performed in this set of experiments have examined the effectiveness of silver-exchange zeolite and argentite and the results of these experiments are presented in Figure 2. For silver zeolite, the removal of the iodide species from solution occurred largely in the first hours of contact. However, the iodine concentrations given in the figure are below the detection limit of 25 ppm for the ICP-MS in the high-Na solution so observations of more detailed trends are not possible. For the batch sorption experiments in DIW, the iodine was removed quickly from solution but the concentration seemed to fluctuate between 0.01 and 0.1 ppm indicating that other processes might have an effect on the aqueous I concentrations in these experiments.

Results for argentite show a slightly different trend where iodide removal was slower than that observed in the experiments conducted with the silver zeolite and the removal of iodide from the 7.8 M Na Ave solution was lower than when the experiment was performed in DIW. As with the BFS situation, this may be a result of kinetic limitations of argentite where the mineral is slowing dissolving and releasing available Ag(I) that can react with iodide.

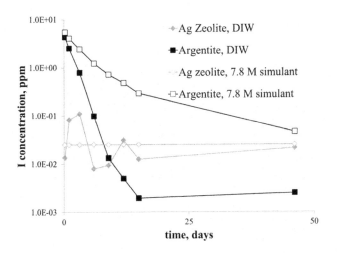

Figure 2. The removal of iodide from solution as a function of time for Ag Zeolite and Argentite. Results are presented for experiments performed in DIW and the 7.8 M Na Ave simulant with Cr. The initial iodide spike is approximately 6.5 ppm. The values for Ag zeolite in the 7.8 M simulant correspond to the detection limit.

Solid Phase Characterization for Tc getters

SEM was used to examine the surface morphology while EDS was used to investigate the distribution of Tc on the surface of the various Tc getter materials. However, only the SEM images from samples contacting DIW were useful because of the large Na interference originating from the 7.8 M Na Ave simulant. Two images obtained with Tin Apatite and BFS 2 are presented in Figure 3. We note that the EDS results are qualitative in nature so they only can be used as indicators of the general location of Tc on the sample. The figure for Tin Apatite {theoretical formula $[Ca_5(PO_4)_3(OH,F,Cl)]$} shows that Tc was evenly distributed across the sample surface. However, there does not seem to be a distinct morphology associated with the presence of Tc which would aid in identifying any secondary phase responsible for Tc sequestration. For the BSF 2 image and corresponding spectra, the Tc seemed to be sequestered into an unidentified phase and not evenly distributed across the surface. A previous study has identified the phase in BFS with the highest reductive capacity as a CaS phase (Oldhamite) and this may be the Tc-rich phase located in the micrograph [14].

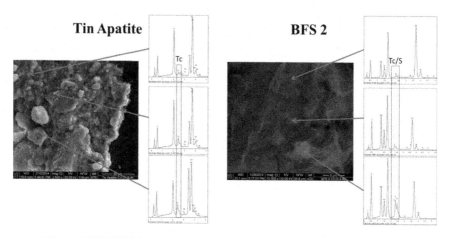

Figure 3. SEM/EDS images for a Tin Apatite (left) and BFS 2 (right). Samples were obtained from batch sorption experiments of Tc by getters in DIW.

CONCLUSIONS

The data presented here show that there are many possible getter materials available that remove Tc and I from DIW during short-term tests in an anoxic chamber. The removal process for Tc is most likely a reduction from the Tc(VII) species in solution to the less soluble Tc(IV). This reduction is thought to occur at the getter surface. The removal of iodide from solution is most likely due to the precipitation of AgI although we have not obtained evidence this from post-reaction solid-state analysis to confirm this. Results from the series of batch sorption experiments performed in the 7.8 M Na Ave LAW simulant indicate that the getter materials may be less effective in this medium. For Tc getters, the limited effectiveness in the LAW simulant seems to be a result of Cr competition and possibly a pH effect. In the high-pH conditions a negative charge may develop on the surface that may repel negatively-charged ions from the surface. For the I getters, the limited process may be exposure to Ag^+ ions in solution that can react with I^- to form the largely insoluble AgI. Future studies will also investigate the difference in removal of another aqueous iodine species, IO_3^-, to see if I getters can effectively remove this species from solution.

Though some of the getters show promising results, the use of the material must be proven as useful in enhancing Tc and I retention in the final Cast Stone waste form in both short-term and long-term waste form leach tests. Further tests are in progress to investigate this. In one of the tests, the getter material is allowed to remove the radionuclides from solution and then the rest of the Cast Stone dry blend ingredients are added and the product is allowed to cure. In another test, all of the dry blend ingredients and getter are added to the liquid waste simulant and the cementitious material is allowed to cure. Both of these resulting materials will undergo performance testing using standard leaching protocols. Results from these studies should indicate if getter use in Cast Stone is a viable option to minimize ^{99}Tc and ^{129}I release to the environment.

ACKNOWLEDGMENTS

This work was completed as part of the Supplemental Immobilization of Hanford Low-Activity Waste and Technetium Management-Hanford site projects. Support for this project came from the U.S. Department of Energy Office of Environment Management. The authors wish to acknowledge Dave Swanberg (Washington River Protection Solutions, Supplemental Treatment Waste Form Development Project) for programmatic guidance, direction, and support and for synthetizing and providing the tin apatite material used in this study. The authors would also like to acknowledge Ian Leavy and Steven Baum for their help in running the ICP-MS. The authors wish to thank Guohui Wang of PNNL for his technical peer review. The SEM/EDS described in this paper was performed in part in the Environmental Molecular Sciences Laboratory, a national scientific user facility sponsored by the U.S. Department of Energy's Office of Biological and Environmental Research and located at Pacific Northwest National Laboratory in Richland, WA.

REFERENCES

[1] Langton, CA, Dukes, MD, Simmons, RV (1984) Cement-Based Waste Forms for Disposal of Savannah River Plant Low-Level Radioactive Salt Waste. In: G.L. McVay (Ed.), *Scientific Basis for Nuclear Waste Management VII*. Materials Research Society, Boston, MA, **26**.

[2] Westsik Jr, JH, Piepel, GF, Lindberg, MJ, Heasler, PG, Mercier, TM, Russell, RL, Cozzi, AD, Daniel, WE, Eibling, RE, Hansen, EK, Reigal, MR, Swanberg, DJ (2013) Supplemental Immobilization of Hanford Low-Activity Waste: Cast Stone Screening Tests. PNNL-22747, SRNL-STI-2013-00465, Rev. 0, Pacific Northwest National Laboratory, Richland, WA and Savannah River National Laboratory, Aiken, SC.

[3] Chung, CW, Turo, LA, Ryan, JV, Johnson, BR, McCloy, JS (2013) The effect of concentration on the structure and crystallinity of a cementitious waste form for caustic wastes. *Journal of Nuclear Materials*, **437**, 332-340.

[4] Mann, FM, Puigh, RJ, Finfrock, SH, Khaleel, R, Wood, MI (2003) Integrated Disposal Facility Risk Assessment. RPP-15834, Rev. 0, CH2M Hill Hanford Group, Inc., Richland, WA.

[5] DOE (2012) Final Tank Closure and Waste Management Environmental Impact Statement for the Hanford Site, Richland, Washington (TC & WM EIS). DOE/EIS-0391, US Department of Energy, Office of River Protection, Richland, WA.

[6] Pierce, EM, Mattigod, SV, Westsik Jr, JH, Serne, RJ, Icenhower, JP, Scheele, RD, Um, W, Qafoku, NP (2010) Review of Potential Candidate Stabilization Technologies for Liquid and Solid Secondary Waste Streams. PNNL-19122, Pacific Northwest National Laboratory, Richland, WA.

[7] Duncan, JB, Cooke, GA, Lockrem, LL (2008) Assessment of Technetium Leachability in Cement-Stablized Basin 43 Groundwater Brine. RPP-RPT-39195, Rev. 1, US Department of Energy, Richland, WA.

[8] Qafoku, NP, Neeway, JJ, Lawter, AR, Levitskaia, TG, Serne, RJ, Westsik Jr, JH, Valenta Snyder, MM (2014) Technetium and Iodine Getters to Improve Cast Stone Performance. PNNL-23282, Pacific Northwest National Laboratory, Richland, WA.

[9] Neeway, JJ, Qafoku, NP, Serne, RJ, Lawter, AR, Stephenson, JR, Lukens, WW, Westsik Jr, JH (2014) Evaluation of Technetium Getters to Improve the Performance of Cast Stone. PNNL-23667, Pacific Northwest National Laboratory, Richland, WA.

[10] Asmussen, RM, Neeway, JJ, Qafoku, NP (2015) Technetium and Iodine Getters to Improve Cast Stone Performance – 15420, *WM2015 Conference*, Phoenix, AZ, USA.

[11] Um, W, Valenta, MM, Chung, CW, Yang, J, Engelhard, MH, Serne, RJ, Parker, K, Wang, G, Cantrell, KJ, Westsik Jr, JH (2011) Radionuclide Retention Mechanisms in Secondary Waste-Form Testing: Phase II. PNNL-20753, Pacific Northwest National Laboratory, Richland, WA.

[12] Serne, RJ, Westsik Jr, JH (2011) Data Package for Secondary Waste Form Down-Selection—Cast Stone. PNNL-20706, Pacific Northwest National Laboratory, Richland, WA.

[13] Kaplan, DI, Mattigod, SV, Parker, KE, Iversen, G (2000) Experimental Work in Support of the [129]I Disposal Special Analysis. WSRC-TR-2000-00283, Rev. 0, Westinghouse Savannah River Company, Aiken, SC.

[14] Um, W, Jung, HB, Wang, G, Westsik Jr, JH, Peterson, RA (2013) Characterization of Technetium Speciation in Cast Stone. PNNL-22977, Pacific Northwest National Laboratory, Richland, WA.

Mater. Res. Soc. Symp. Proc. Vol. 1744 © 2015 Materials Research Society
DOI: 10.1557/opl.2015.298

Selective Ordering of Pertechnetate at the Interface between Amorphous Silica and Water: a Poisson Boltzmann Treatment

Christopher D. Williams[1,2], Karl P. Travis[1], John H. Harding[1], Neil A. Burton[2]
[1]Department of Materials Science and Engineering, University of Sheffield, Sheffield, S1 3JD, U.K. [2]School of Chemistry, University of Manchester, Manchester, M13 9PL, U.K.

ABSTRACT

Calculations based on Poisson-Boltzmann theory are used to investigate the equilibrium properties of an electrolyte containing TcO_4^- and SO_4^{2-} ions near the surface of amorphous silica. The calculations show that the concentration of TcO_4^- is greater than SO_4^{2-} at distances less than 1 nm from the surface due to the negative charge density caused by deprotonation of the amorphous silica silanol groups. At lower pH, the surface becomes protonated and the magnitude of this effect is reduced. These results have implications for the potential use of oxyanion-SAMMS for the environmental remediation of water contaminated with ^{99}Tc.

INTRODUCTION

The environmental remediation of ^{99}Tc contaminated land and water is one of the most difficult challenges facing the nuclear industry [1]. The problem stems from the fact that ^{99}Tc has a long half-life (2.1×10^5 years), high fission yield and readily forms the extremely mobile pertechnetate oxyanion, TcO_4^-, in aqueous solution [2]. Many of the previously tested methods of TcO_4^- remediation suffer from poor selectivity in the presence of competing anions, such as sulphate (SO_4^{2-}), or are dependent on specific reducing conditions [3]. However, self-assembled monolayers on mesoporous supports (SAMMS) have been proposed as an effective alternative for the difficult environmental remediation problems [4]. SAMMS combine a robust amorphous silica support, known as MCM-41 [5], with a highly selective functionalized monolayer that can be tuned to target the contaminant species of interest. Experiments have shown that SAMMS and related materials can adsorb monovalent anions, such as TcO_4^-, preferentially over the competing divalent SO_4^{2-} [6, 7]. In order to tailor the design of oxyanion-SAMMS to optimize their performance, an improved understanding of the cause of this selectivity is required.

Amorphous silica surfaces have a high density of silanol groups [8]. If the pH at the point of zero charge of the surface, pH_{pzc}, is exceeded by the pH of the wetting solution there will be a net negative charge due to deprotonation of these silanols. Since the pH_{pzc} of amorphous silica is very low ($2 - 3.5$ [9]) the surface will be negatively charged in the range of pH encountered in groundwater and the extent of ionization will determine the electrostatic potential near the surface. Helmholtz proposed that if a charged surface is placed in contact with an electrolyte then a neutralizing electrical double layer of counterions forms at the interface [10]. Here, Poisson-Boltzmann (PB) theory is used to calculate the equilibrium properties of an electrolyte containing the two competing TcO_4^- and SO_4^{2-} in contact with an amorphous silica surface.

THEORY

The equilibrium properties of a system of interacting ions near a planar charged surface, positioned at $z = 0$ and extending infinitely in the x and y directions, was treated using the one-dimensional PB equation

$$\frac{d^2\phi_{elec}}{dz^2} = \frac{e}{\varepsilon_0\varepsilon_r}\sum_i^N q_i [X_i] \exp\left[\frac{q_ie\phi_{elec}(z)}{k_BT}\right] \tag{1}$$

where e, ε_0, ε_r, k_B and T are the elementary charge, vacuum permittivity, relative permittivity of the solvent, Boltzmann constant and temperature, respectively. N is the total number of ionic species with charge q_i and bulk concentration $[X_i]$. $\phi_{elec}(z)$ is the mean electrostatic potential at a distance z normal to the surface. The central tenet of PB theory is the mean-field approximation, which means that the correlated positions of individual ions in the system are neglected and only their mean equilibrium distributions are considered. In addition, only coulombic interactions between ions, which are modeled as point-like objects, are accounted for. The ions are considered to be mobile within a solvent that is modeled as a continuous medium with a fixed dielectric constant. Analytical solutions to the PB equation, such as the Gouy-Chapman-Stern model [10], are only possible for very simple electrolytes. For more complex electrolytes, such as those encountered in the aqueous environment, the PB equation must be solved numerically.

The calculation described here follows the methodology of Lochhead et $al.$ [11] and was implemented by adapting a program originally developed to investigate the electrostatic properties of a buffer solution at the surface of cantilever arrays used as sensors [12]. The model electrolyte was considered to be a mix of the fully dissociated potassium salts of the two oxyanions (KTcO$_4$ and K$_2$SO$_4$) in water ($\varepsilon_r = 78.0$) at 298 K. Chemical equilibria for the dissociation of water and subsequent protonation of the oxyanions must be considered to obtain realistic estimates of the bulk concentrations of ions in the electrolyte using the pK_a values for HTcO$_4$ (0.033), H$_2$SO$_4$ (−2) and HSO$_4^-$ (1.99) [10, 13]. The electrolyte therefore consists of H$^+$, OH$^-$, K$^+$, TcO$_4^-$, SO$_4^{2-}$ and HSO$_4^-$. Since all the equilibria involve [H$^+$] they are not independent of each other and the bulk concentrations must be solved self-consistently, under the conditions of fixed ionic strength and pH.

The surface was modeled as a uniformly smeared charged density, proportional to the concentration of fully dissociated silanol groups on the amorphous silica surface ([SiO$^-$] = 5.0 nm^{-2}) [14]. Cations in the electrical double layer are in equilibrium with those chemically bound to the surface, where the extent of binding is governed by the formation constant

$$K_M = \frac{[SiOM]}{[SiO^-][M^+]_0} \tag{2}$$

where [M$^+$]$_0$ is the concentration of cations at the surface. The equilibrium constants introduce chemical specificity into the ionic interactions with the surface, which are otherwise indistinguishable in PB theory. The binding of cations affects the surface charge density, which was obtained from the equation [11]

$$\sigma = \frac{-e[SiO^-]}{(1+K_H[H^+]_0+K_K[K^+]_0)} \tag{3}$$

where the local concentration of ions at a distance z from the surface can be obtained from the Boltzmann distribution

$$[X_i]_z = [X_i]\exp\left[\frac{-q_ie\phi_{elec}(z)}{k_BT}\right] \tag{4}$$

54

and values for K_H and K_K of 10^6 and 0.3162 dm^3 mol^{-1} were used [15, 16]. The electrostatic potential at the surface was then calculated from the surface charge density using the Grahame equation [17]

$$\frac{\sigma}{\varepsilon_r \varepsilon_0} = -\sqrt{\frac{2k_B T}{\varepsilon_r \varepsilon_0} \sum_i^N [X_i] \left[\exp\left[\frac{-q_i e \phi_{elec}(z)}{k_B T}\right] - 1\right]} \tag{5}$$

$\phi_{elec}(0)$, σ and $[X_i]_0$ were determined by solving Equations 2, 3 and 5 simultaneously, under the condition that the total charge of ions in solution must be equal and opposite to the charge on the surface. Using the electrostatic potential at the surface the full potential profile was calculated by numerical integration of Equation 1. The mean concentration profiles for the ions, which are assumed to be in thermodynamic equilibrium having adjusted to the electrostatic potential profile, were then obtained using Equation 4.

The formation of an electrical double layer leads to high concentrations of ions at the surface and large deviations from ideal behavior for each ionic species. The Debye-Huckel and Truesdell-Jones models are not valid at these concentrations so the Pitzer model [18], a virial coefficient approach that consists of linear combinations of empirical parameters, was instead used to determine single-ion activity coefficients. The parameters used in the Pitzer model were taken from Neck *et al.* [19]. Since the activity coefficient of one ionic species is dependent on the concentrations of all the others, which are unknown prior to the start of the calculation, activity coefficients were determined iteratively, whereby they are set to unity the first time the Grahame equation was solved.

RESULTS AND DISCUSSION

In a groundwater remediation scenario the contaminant species are typically present at trace concentration compared to environmentally ubiquitous ions; we therefore considered a model electrolyte that contains trace concentrations of TcO_4^- (10^{-10} mol dm^{-3}) relative to SO_4^{2-} (10^{-3} mol dm^{-3}), at pH 7. The electrostatic potential and ion concentration profiles near the surface were determined for this electrolyte. The percentage of sites bound by H^+ and K^+ were 26% and 53%, respectively, leaving 21% ionized sites. ϕ_{elec} is most negative at the surface (-274 mV), decreasing slowly, due to a combination of the high surface charge density and low solution ionic strength, to -114 mV at 4.0 nm. The concentration profiles of the ions are shown in Figure 1. HSO_4^- is omitted as it is present only in very minor concentrations compared to the fully deprotonated ion at this pH. The increase in H^+ and K^+ concentrations near the surface relative to the bulk indicate the formation of an electrical double layer. Conversely, the concentration of anions is depleted near the surface, due to electrostatic repulsion. Far from the surface the concentrations of the two oxyanions tend toward their bulk values ($[SO_4^{2-}] >$ $[TcO_4^-]$) but at distances less than 1.0 nm from the surface this trend is reversed ($[TcO_4^-] >$ $[SO_4^{2-}]$). At the surface the concentration of TcO_4^- is higher ($[TcO_4^-]_0 = 3.1 \times 10^{-14}$ mol dm^{-3} vs. $[SO_4^{2-}]_0 = 5.9 \times 10^{-17}$ mol dm^{-3}), despite the concentration of SO_4^{2-} being seven orders of magnitude greater in bulk solution. This is due to greater electrostatic repulsion caused by the exponential dependence on q_i in Equation 1.

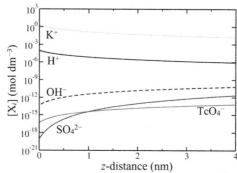

Figure 1. Ionic concentration profiles for a solution containing 10^{-3} mol dm^{-3} of K$_2$SO$_4$ and 10^{-10} mol dm^{-3} of KTcO$_4$ at bulk pH 7 in contact with an amorphous silica surface. The concentrations of H$^+$ (solid) and OH$^-$ (dashed) ions are shown with a black line and K$^+$, TcO$_4^-$ and SO$_4^{2-}$ are shown with green, red and blue lines, respectively.

The pH of aqueous solutions encountered in the environment span a wide range, so the effect of varying the bulk pH of our model electrolyte was investigated. The effect pH has on the electrostatic potential profiles using an electrolyte with same initial bulk oxyanion concentrations is shown in Figure 2. As the bulk solution pH increases, the H$^+$ ion concentration at the surface decreases, resulting in a greater percentage of ionized surface sites and a more negative electrostatic potential near the surface. At bulk pH 1 only 0.2% of surface silanols are ionized and the electrostatic potential at the surface is consequently much lower (−15.2 mV) compared to the surface in contact with a solution with bulk pH 7 (−274 mV). However, calculations performed at a bulk pH below that the pH$_{pzc}$ of the surface (2 − 3.5) are unrealistic since at very low pH the surface becomes fully protonated and is positively charged. These calculations only account for deprotonation of the surface, so they can only be deemed useful for electrolytes with a pH several units greater than the pH$_{pzc}$ of the surface.

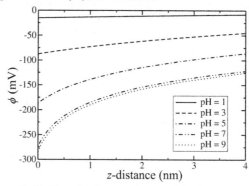

Figure 2. Effect of bulk pH on the electrostatic potential near the charged amorphous silica surface.

The effect of bulk pH on the concentrations of oxyanions 1.0 nm from the surface is shown in Figure 3. The values obtained at pH 7 correspond to the crossover in the TcO_4^- and SO_4^{2-} concentrations shown previously in Figure 1. Table 1 shows the surface potential and the resulting concentrations of oxyanions at 1.0 nm for solutions with pH 4 – 9. At low pH the concentration of SO_4^{2-} was found to be several orders of magnitude greater than TcO_4^-. As the magnitude of the surface charge increases with bulk pH electrostatic repulsion from the surface increases. This effect has a more significant impact on the SO_4^{2-} concentration profile. At pH 7 the concentrations of the two oxyanions are very similar and as the bulk pH is increased further TcO_4^- becomes the dominant species. Above pH 9 there is little change in the relative concentrations as further decreases in $[H^+]$ do not significantly affect the percentage of ionized surface sites as K^+ is the dominant bound cation. If the model was adapted to account for a fully protonated surface the concentration of oxyanions near the surface would actually be greater than in bulk solution. The electrical double layer would then be dominated by SO_4^{2-} due to its higher charge than the other anions in this system (OH^- and TcO_4^-).

Table 1. The effect of bulk pH on the extent of surface ionisation, electrostatic potential and oxyanion concentration 1.0 nm from the surface.

pH	Ionised Surface Sites (%)	ϕ_{elec} (mV)	$[TcO_4^-]$ (mol dm^{-3})	$[SO_4^{2-}]$ (mol dm^{-3})
4	3.2	−106.7	5.52×10^{-13}	3.81×10^{-9}
5	6.8	−138.9	8.34×10^{-14}	2.41×10^{-11}
6	13.5	−166.9	1.33×10^{-14}	1.37×10^{-13}
7	20.7	−184.2	3.96×10^{-15}	4.19×10^{-15}
8	23.1	−187.5	2.77×10^{-15}	1.52×10^{-15}
9	23.4	−188.6	2.60×10^{-15}	1.28×10^{-15}

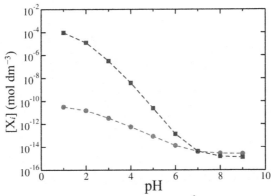

Figure 3. Dependence of $[TcO_4^-]$ (red, circles) and $[SO_4^{2-}]$ (blue, squares) on bulk solution pH, 1.0 nm from the surface.

CONCLUSIONS

The properties of an electrolyte in contact with an amorphous silica surface were investigated using PB theory. Oxyanions were repelled from the surface due to its negative charge density when in contact with a solution with a pH higher than the pH_{pzc}. Repulsion of the divalent SO_4^{2-} anion is greater than the monovalent TcO_4^-. As a result, at pH 7, the concentration of TcO_4^- near the surface is significantly higher than SO_4^{2-} even when its concentration in bulk solution is seven orders of magnitude lower. In SAMMS, although the majority of surface silanols on the internal pore surface have been functionalized, the effects investigated here may influence the entry of oxyanions into the pores.

ACKNOWLEDGMENTS

We thank the Engineering and Physical Sciences Research Council and Nuclear FiRST Centre for Doctoral Training for funding.

REFERENCES

1. *Generic Repository Studies: Generic post-closure Performance Assessment Report number N/080*. 2003, UK Nirex Limited.
2. K.H. Lieser and C. Bauscher, Radiochim. Acta **42** (4), 205, (1987).
3. E.H. Schulte and P. Scoppa, Sci. Total Environ. **64** (1-2), 163, (1987).
4. X.B. Chen, X.D. Feng, J. Liu, G.E. Fryxell, and M.L. Gong, Sep. Sci. Technol. **34** (6-7), 1121, (1999).
5. C.T. Kresge, M.E. Leonowicz, W.J. Roth, J.C. Vartuli, and J.S. Beck, Nature **359** (6397), 710, (1992).
6. H. Yoshitake, T. Yokoi, and T. Tatsumi, Chem. Mater. **15** (8), 1713, (2003).
7. G.E. Fryxell, J. Liu, T.A. Hauser, Z.M. Nie, K.F. Ferris, S. Mattigod, M.L. Gong, and R.T. Hallen, Chem. Mater. **11** (8), 2148, (1999).
8. H. Landmesser, H. Kosslick, W. Storek, and R. Fricke, Solid State Ionics **101**, 271, (1997).
9. M. Kosmulski, J. Colloid Interface Sci. **253** (1), 77, (2002).
10. P. Atkins and J. De Paula, *Physical Chemistry*, 9th ed. (University Press, Oxford, 2006), p. 861.
11. M.J. Lochhead, S.R. Letellier, and V. Vogel, J. Phys. Chem. B **101** (50), 10821, (1997).
12. M.L. Sushko, J.H. Harding, A.L. Shluger, R.A. McKendry, and M. Watari, Adv. Mater. **20** (20), 3848, (2008).
13. H. Stephan, K. Gloe, W. Kraus, H. Spies, B. Johannsen, K. Wichmann, G. Reck, D.K. Chand, P.K. Bharadwaj, U. Muller, W.M. Muller, and F. Vogtle, in *Anion Separations: Fundamentals and Applications*, edited by R.P. Singh and B.A. Moyer (Kluwer, New York, 2003) p.151.
14. L.T. Zhuravlev, Langmuir **3** (3), 316, (1987).
15. A. Carre and V. Lacarriere, Contact Angle, Wettability and Adhesion **4**, 267, (2006).
16. M. Berka and I. Banyai, J. Colloid Interface Sci. **233** (1), 131, (2001).
17. D.C. Grahame, J. Chem. Phys. **21** (6), 1054, (1953).
18. K.S. Pitzer, in *Ion interaction approach: theory and data correlation*, edited by K.S. Pitzer (CRC Press, Boca Raton, 1991) p. 279.
19. V. Neck, T. Konnecke, T. Fanghanel, and J.I. Kim, J. Solution Chem. **27** (2), 107, (1998).

Development and Characterization of Waste Forms

Mater. Res. Soc. Symp. Proc. Vol. 1744 © 2015 Materials Research Society
DOI: 10.1557/opl.2015.394

Pressureless Sintering of Sodalite Waste-forms for the Immobilization of Pyroprocessing Wastes

M. R. Gilbert
AWE, Aldermaston, Reading, RG7 4PR, UK.

ABSTRACT

Sodalite ($Na_8[AlSiO_4]_6Cl_2$), a naturally occurring Cl-containing mineral, has long been regarded as a potential immobilization matrix for the chloride salt wastes arising from pyrochemical reprocessing operations, as it allows for the conditioning of the waste salt as a whole without the need for any pre-treatment. Here the consolidation and densification of Sm-doped sodalite (as an analogue for $AnCl_3$) has been investigated with the aim of producing fully dense (i.e. > 95 % t.d.) ceramic monoliths via conventional cold-press-and-sinter techniques at temperatures of < 1000 °C. Microstructural analysis of pressed and sintered sodalite powders under these conditions is shown to produce poorly sintered, porous, inhomogeneous pellets. However, by the addition of a sodium aluminophosphate glass sintering aid, fully dense Sm-sodalite ceramic monoliths can successfully be produced by sintering at temperatures as low as 800 °C.

INTRODUCTION

Pyrochemical reprocessing techniques enable the recovery of Pu metal from spent nuclear material without the need to convert it to PuO_2 and back [1]. These methods utilise an electrorefining process, where the Pu is separated from the impurities in a molten chloride salt, most typically either $CaCl_2$ or an equimolar mixture of NaCl-KCl, at temperatures of between 750 – 850 °C [2]. Post-reprocessing, this chloride salt must be replaced, as it now contains a number of different waste streams which will contaminate the cathode and affect the properties of the molten salt. This contaminated salt must be disposed of in such a way as to immobilize the radionuclide chlorides contained within. However, halide-rich wastes such as these can be problematic to immobilize, as not only are their solubilities in melts very low, but even in small quantities they can seriously affect the properties of the waste-form [3,4]. In addition, processing temperatures are often severely limited in order to prevent the volatilisation of the halides.

One approach is to immobilize the radionuclide chlorides in a ceramic waste-form based upon mineral phases with naturally high chloride contents, of which sodalite is one such example. Natural sodalite, $Na_8[AlSiO_4]_6Cl_2$, is a naturally chlorine-containing mineral, containing up to 7.3 wt. % Cl [5]. It is a crystalline aluminosilicate formed of a framework of all-corner-linked tetrahedra. Sodalite is formed by the fusing of the 4-membered rings of the β-cages, such that its framework structure consists solely of β-cages, creating a microporous structure with a typical pore diameter of 4 Å [6]. Work carried out at Argonne National Laboratory on glass-bonded sodalites has demonstrated them to be effective for immobilizing wastes from NaCl-KCl based reprocessing, though preparation temperatures range from 850 – 1000 °C [7-9]. In addition, work at Idaho National Laboratory formed sodalite waste-forms by first incorporating the molten salt into zeolite-4A before then hot pressing at ~ 850 °C in order to convert the zeolite to sodalite [10,11].

More recently, B. J. Riley and co-workers at Pacific Northwest National Laboratory have developed a solution-based synthesis method for the formation of sodalite waste-forms [12]. To improve consolidation and densification the group has utilized a number of different glass sintering aids in order to produce sintered waste-forms both by conventional cold-press-and-sinter and hot pressing techniques, demonstrating potential theoretical densities of up to 92 % [13,14].

EXPERIMENT

Solid solutions of $(Na_{8-3x}Sm_x)(AlSiO_4)_6Cl_2$, where $x = 0$, 0.05, 0.1 and 0.2, were prepared via an anhydrous nepheline $(NaAlSiO_4)$ intermediate (Equations 1 and 2) based on the method described by De Angelis et al [15].

$$Al_2Si_2O_7.2H_2O + 2NaOH \rightarrow 2NaAlSiO_4 + 3H_2O \qquad (1)$$

$$6NaAlSiO_4 + (2-3x)NaCl + xSmCl_3 \rightarrow (Na_{8-3x}Sm_x)(AlSiO_4)_6Cl_2 \qquad (2)$$

Nepheline was first synthesized by weighing stoichiometric quantities of kaolinite $(Al_2Si_2O_7.2H_2O)$ and NaOH, together with a 5 wt. % excess of NaOH, into a Nalgene mill pot with YSZ milling media and dry milling overnight. The resulting powder mix was passed through a 250 μm sieve mesh, placed in an alumina crucible, heated to 300 °C to drive off any absorbed moisture and then calcined at 700 °C for 3 h in air. The calcined nepheline plug was then broken up using a pestle and mortar and passed through a 212 μm sieve mesh.

Stoichiometric quantities of $NaAlSiO_4$, NaCl and $SmCl_3$ were then placed into a Nalgene mill pot together with YSZ milling media and dry milled overnight. The resulting powder mix was passed through a 250 μm sieve mesh, placed in an alumina crucible, heated to 300 °C to drive off any absorbed moisture and then calcined at 750 °C for 48 h in air to produce the Sm-doped sodalites. Pellets were uniaxially pressed, both with and without glass sintering aid, at pressures of 80 – 200 MPa and then sintered in air at temperatures of 800 – 1000 °C for 12 h. Pellet densities were measured using the Archimedes method.

Powder XRD was carried out using a Bruker D8 Advance diffractometer operating in Bragg-Brantano flat plane geometry using Cu $K_{\alpha1}$ radiation ($\lambda = 1.54056$ Å). Diffraction patterns were measured over a 2θ range of 10 – 90° using a step size of 0.025° and a collection time of 3.4 s per step.

The microstructures of the sintered pellets were studied using scanning electron microscopy (SEM) and energy dispersive X-ray analysis (EDX). A Hitachi TM-1000 electron microscope was used with a 15 kV accelerating voltage in backscattering mode. A Bruker AXS QUANTAX 50 EDAX system was used for elemental analysis of the samples.

RESULTS

XRD patterns of both undoped and Sm-doped sodalite are shown in Figure 1. The XRD pattern of the undoped $Na_8(AlSiO_4)_6Cl_2$ shows that single-phase sodalite has been successfully fabricated. Analysis of the Sm-doped sodalite patterns shows the $(Na_{8-3x}Sm_x)(AlSiO_4)_6Cl_2$

compositions remain almost entirely single-phase sodalite. The first indications of a minor phase(s) can be seen appearing in the x = 0.1 and x = 0.2 compositions in the region 21 – 30° 2θ, however, at this current level of doping the intensity is too low for these to be identified.

Initially, pellets of pure sodalite (without the addition of the glass sintering aid) were uniaxially pressed at pressures of 100 – 200 MPa and sintered in air at temperatures of 800, 900 and 1000 °C for 12 h. However, densification was found to be very poor across the whole range. Whilst small areas of well sintered material could be observed, the majority of the pellet microstructures showed poorly sintered, porous, inhomogeneous pellets, as can be seen in Figure 2, which shows a comparison of 140 MPa sodalite pellets sintered at 800, 900 and 1000 °C.

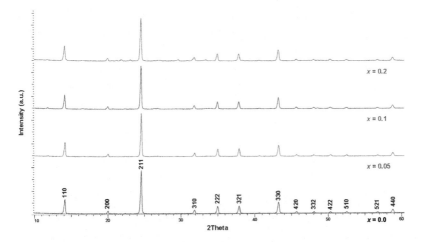

Figure 1. XRD patterns of $(Na_{8-3x}Sm_x)(AlSiO_4)_6Cl_2$ calcined at 750 °C for 48 h in air.

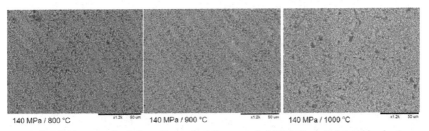

140 MPa / 800 °C   140 MPa / 900 °C   140 MPa / 1000 °C

Figure 2. BSE images of sodalite pellets uniaxially pressed at 140 MPa and sintered in air at 800, 900 and 1000 °C for 12 h.

Following this, an aluminophosphate glass sintering aid, known as GTI/168, was added, the composition of which is shown in Table 1. 25 wt. % GTI/168 was added to each sodalite pellet and uniaxially pressed at pressures of 80 – 200 MPa before being sintered in air at 800 °C for 12 h. The measured densities are shown in Table 2 together with the % theoretical density (t.d.), and backscattered electron (BSE) images of the pellet surfaces are shown in Figure 3.

Table 1. GTI/168 composition.

GTI/168	wt. %	mol. %
Na_2O	24.65	40.1
P_2O_5	54.62	38.8
Al_2O_3	19.33	19.1
B_2O_3	1.40	2.0

Table 2. Measured densities for 75:25 Sodalite:GTI/168 pellets sintered in air at 800 °C for 12 h.

Applied Pressure / MPa	Density / $g.cm^{-3}$	% t.d.
80	2.01	96.0
100	2.02	96.3
120	2.00	95.1
140	1.92	91.2
160	1.93	91.8
180	1.89	89.8
200	1.85	88.2

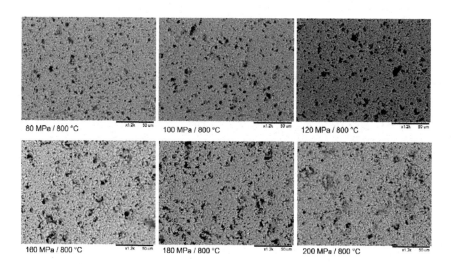

Figure 3. BSE images of 75:25 Sodalite:GTI/168 pellets uniaxially pressed at 80 – 200 MPa and sintered in air at 800 °C for 12 h.

Whilst EDX analysis of the pellet surfaces clearly shows the presence of Na, Al and P from the GTI/168, there is a notable absence of any distinct glassy phase in the pellet microstructures. Instead, small quantities of a secondary phase, darker in contrast than the sodalite can now be seen. Subsequent XRD analysis of the crushed pellets (Figure 4) shows very little amorphous contribution from any glassy matrix, but instead a number of new diffraction peaks from $20 - 32$ $^\circ 2\theta$ (Figure 4 inset). These relate to aluminium and silicon phosphates, and a number of sodium aluminosilicate phases, all of which have arisen from the reaction of the GTI/168 glass sintering aid with the sodalite itself. Ultimately however, they are very minor secondary phases, all pellets remaining near single-phase sodalite.

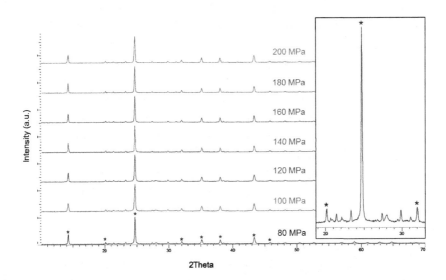

Figure 4. XRD patterns of sintered 75:25 Sodalite:GTI/168 pellets. Inset shows new diffractions peaks arising from aluminium phosphate, silicon phosphate and sodium aluminosilicate phases. * denotes sodalite peak.

CONCLUSIONS

The pressureless sintering of pure sodalite has been shown not to produce fully dense (i.e. ≥ 95 % t.d.) pellets, even at pressures of up to 200 MPa and temperatures of up to 1000 $^\circ$C. The resulting pellets display poorly sintered, inhomogeneous microstructures and, although the sintering behaviour improves with increasing pressure and temperature, these conditions never produce anything more than small, isolated regions of well sintered material.

The addition of 25 wt. % GTI/168 sodium aluminophosphate glass as a sintering aid dramatically improves the sintering behaviour of the sodalite. When pressed at 80 – 120 MPa, pellets of > 95 % t.d. are formed from sintering at just 800 °C. At higher pressures the densities begin to drop away as a result of overpressing, although still remain > 88 % t.d., an improvement over those pellets sintered with no GTI/168.

No glassy phase is observed in the sintered 75:25 Sodalite:GTI/168 pellets. Analysis of the crushed pellets by powder XRD shows that, whilst the pellets remain near single-phase sodalite, a number of new aluminium phosphate, silicon phosphate and sodium aluminosilicate crystalline phases have been formed as a result of reaction of the sodium aluminophosphate sintering aid with the sodalite.

REFERENCES

1.   T. Nishimure, T. Koyama, M. Iizuka, H. Tanaka, *Prog. Nucl. Energy,* **32**, 381 (1998).
2.   I. N. Taylor, M. L. Thompson, T. R. Johnson, *Proceedings of the International Conference and Technology Exposition on Future Nuclear,* **1**, 690 (1993).
3.   W. E. Lee, R. W. Grimes, *Energy Materials,* **1**, 22 (2006).
4.   T.O. Sandland, L.-S. Du, J.F. Stebbins, J.D. Webster, *Geochim. Cosmochim. Acta,* **68**, 5059 (2004).
5.   G. Leturcq, A. Grandjean, D. Rigaud, P. Perouty, M. Charlot, *J. Nucl. Mater.,* **347**, 1-11 (2005).
6.   J. Rouguerol, D. Anvir, C. W. Fairbridge, D. H. Everett, J. H. Haynes, N. Pernicone, J. D. Ramsay, K. S. W. Sing, K. K. Unger, *Pure Appl. Chem.,* **66**, 1739 (1994).
7.   M. A. Lewis, C. Pereira, US Patent No. 5,613,240, (18 Mar 1997).
8.   C. Pereira, ANL/CMT/CP--84675, (1996).
9.   S. Priebe, *Nucl. Tech.,* **162**, 199 (2008).
10. M. A. Lewis, D. F. Fischer, L. J. Smith, *J. Am. Ceram. Soc.,* **76**, 2826 (1993).
11. T. J. Moschetti, S. G. Johnson, T. DiSanto, M. H. Noy, A. R. Warren, W. Sinkler, K. M. Goff, K. J. Bateman, *Mater. Res. Soc. Symp. Proc.,* **713**, 329 (2002).
12. B. J. Riley, J. V. Crum, J. Matyáš, J. S. McCloy, W. C. Lepry, *J. Am. Ceram. Soc.,* **95**, 3115 (2012).
13. W. C. Lepry, B. J. Riley, J. V. Crum, C. P. Rodriguez, D. A. Pierce, *J. Nucl. Mater.,* **442**, 350-359 (2013).
14. B. J. Riley, D. A. Pierce, S. M. Frank, J. Matyáš, C. A. Burns, *J. Nucl. Mater.,* **459**, 313-322 (2015).
15. G. De Angelis, I. Bardez-Giboire, M. Mariani, M. Capone, M. Chartier, E. Macerata, *Mater. Res. Soc. Symp. Proc.,* **1193**, 73-78 (2009).

Mater. Res. Soc. Symp. Proc. Vol. 1744 © 2015 Materials Research Society
DOI: 10.1557/opl.2015.330

MoO₃ incorporation in alkaline earth aluminosilicate glasses

Shengheng Tan, Michael I Ojovan, Neil C Hyatt, Russell J Hand

ISL, Department of Materials Science & Engineering, University of Sheffield, Sir Robert

Hadfield Building, Mappin Street, Sheffield, S1 3JD, UK

ABSTRACT

Alkaline earth aluminosilicate glasses (AeAS) with different MoO_3 additions have been produced and assessed. MoO_3 solubility increases with the equimolar substitution of smaller to larger alkaline earths and reaches 5.34 mol% in magnesium aluminosilicate glass (MAS). All visibly homogeneous glasses are X-ray amorphous, while the partially crystallised glasses exhibit some small X-ray diffraction peaks which are probably due to corresponding molybdates. The addition of MoO_3 decreases glass transition and crystallisation temperatures and creates two broad Raman bands which are assigned to vibrations of MoO_4^{2-} tetrahedra. The intensities of these bands increase along with MoO_3 incorporation until the maximum solubility is reached. Electron microscopy shows that these separated particles are spherical, with sub-micron diameters and are randomly dispersed within glass. The separated phases are formed through liquid-liquid separation and thereafter crystallisation. Overall AeAS glasses look quite promising for molybdate immobilisation with MAS glasses being particularly attractive.

INTRODUCTION

Some of the high level nuclear wastes (HLW) produced in the UK and France contain high concentrations of MoO_3 [1-2], one of the most challenging oxides occurring in the vitrification of radioactive wastestreams on account of its low solubility (≤ 1 wt%) in the conventionally used borosilicate glasses [3]. Excess MoO_3 in nuclear waste glasses can cause the formation of "yellow phase" (a mixture of alkali and alkaline earth molybdates with chromates and sulphates) [4], which is detrimental to HLW vitrification process. Yellow phase not only accelerates the corrosion of the melter but also reduces the performance of the vitrified product [5-6]. Therefore, the waste loading capacity for Mo-rich wastes in vitrification has been restricted to avoid the formation of yellow phase. However, post-operational clean-out (POCO) wastes from Sellafield in the UK and some French wastes (e.g. [2]) contain high levels of MoO_3 hence identifying glass composition(s) with high MoO_3 compatibility is desirable.

To the best of our knowledge, little attention has been paid to MoO_3 dissolution in aluminosilicate glasses. Aluminosilicate glasses have been investigated for nuclear waste immobilisation since the late 1950s in Canada [7]. Although these glasses have good glass formation ability, high chemical durability and thermal stability [8-10], practical application have been limited by the high processing temperatures required and low waste loading capacities. This study assesses the capacities of alkaline earth aluminosilicate glasses (AeAS) to incorporate molybdate, the influence of molybdate incorporation on the glass structure and the phase separation which occurs within glasses with excess molybdate.

EXPERIMENTAL

Glass making

AeAS glasses with nominal molar compositions $45SiO_2$-$10Al_2O_3$-$45AeO$-$xMoO_3$ (AeAS-xM, Ae = Mg, Ca, Sr, Ba or two combined, x = 0-8) were produced using laboratory grade precursors of SiO_2, $Al(OH)_3$, MoO_3 and corresponding alkaline earth carbonates (Mg, Ca, Sr) or hydroxide (Mg). Batches to produce ~50 g glass were placed in mullite crucibles. The glass batches were heated in an electric furnace from room temperature to 1450 °C at 2 °C min^{-1}, held for 3 h and afterwards poured out into a preheated steel mould to form a block. The cast glass was immediately transferred to another electric furnace where it was annealed at 700 °C for 1 h and cooled down to room temperature at 1 °C min^{-1}. All of the above procedures were carried out in air.

Characterization

The obtained glasses were partly sectioned to ~5 mm thick slices using a Buehler low speed saw with a diamond blade and with oil as lubricant. The top surfaces of the slices were polished to 1200 SiC grit, rinsed with isopropanol and thoroughly dried. Other parts of glasses were crushed into pieces and ground to fine powders. Powders passing a 75 μm sieve were collected for use.

X-ray diffraction (XRD) and differential thermal analysis (DTA) were performed on the collected sample powders, with a scanning range of 10 – 60 °2θ, a step size of 0.05 °2θ and dwell time of 7 s. DTA curves were recorded upon heating from room temperature to 1200 °C at 10 °C min^{-1} in air. Raman spectroscopy was performed on the polished glass slices using a green line laser (514.5 nm), energy scanning from 0 to 2000 cm^{-1}. Glass slices were further polished to 1 μm with diamond suspension and coated with carbon for the observation of scanning electron microscopy (SEM). Compositional analysis was measured by an attached energy dispersive X-ray spectrometer (EDX) with cobalt calibration every time.

RESULTS AND DISCUSSION

MoO₃ retention and solubility

Glass melts with excess MoO_3 content phase separated during casting on cooling and the resultant glasses are inhomogeneous. MoO_3 solubility in glass is thus defined as the retained MoO_3 content in the homogeneous glass phase with the highest MoO_3 addition. MoO_3 solubilities in all AeAS glasses are high and Figure 1(a) shows that MoO_3 solubility steadily increases with the equimolar substitution of smaller for larger alkaline earths, from 1.85 mol% in BAS glass to 5.34 mol% in MAS glass. The correlation of measured MoO_3 content with batched MoO_3 content is plotted in Figure 1(b). Essentially complete retention of MoO_3 is seen in all glasses except MAS even after the glass composition becomes phase separated (*e.g.* BAS-2.5M and CAS-4M). Essentially complete MoO_3 retention occurs in MAS glass until MAS-6M glass after which the MoO_3 content in glass does not increase anymore. It is likely that MoO_3 incorporation limit has been reached between MAS-6M and MAS-7M glasses and further MoO_3 additions are not likely to enter the glass network.

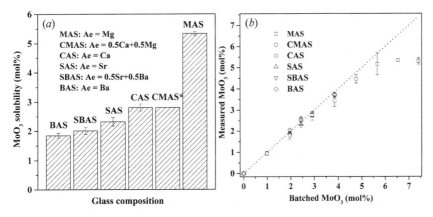

Figure 1 (*a*) MoO₃ solubility and (*b*) MoO₃ retention in different AeAS glasses. *MoO₃ solubility in CMAS glass was not determined; CMAS-4M glass is partially phase separated in a similar manner to CAS-4M glass, thus it is assumed that CAS and CMAS glasses have similar MoO₃ solubilities.

XRD

Figure 2 shows XRD patterns of selected glasses. The patterns for homogeneous glasses display broad humps typical of amorphous materials whereas phase separated glasses display some small crystalline peaks superimposed on these. It is difficult to accurately assign these crystalline peaks, but the highest peaks for each glass are all in good accordance with the patterns of corresponding alkaline earth molybdate crystals. It therefore can be concluded that the separated phases are probably alkaline earth molybdates.

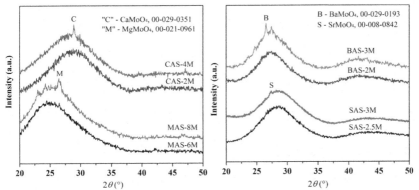

Figure 2 XRD patterns of some prepared glasses with and without phase separation. PDF card numbers are from the PDF4 (2013) database (ICDD).

Raman spectroscopy

Baseline subtraction, frequency correction and normalisation as described in [11] were applied to the Raman spectra. The incorporation of MoO_4^{2-} ions in the glass network results in creation of two broad bands other than the bands assigned to the aluminosilicate network. As seen in Figure 3, the MoO_4^{2-} bands for MAS glass are located at 300-400 cm^{-1} and centred at ~960 cm^{-1}, respectively. The intense 960 cm^{-1} band moves to 920 cm^{-1} for CAS glass due to the Ca^{2+} ions that replace the Mg^{2+} ions associating with the MoO_4^{2-} units. The intensities of MoO_4^{2-} bands are increased with increasing MoO_3 additions in MAS and CAS glasses. However, it can be seen that the increment is gradually reduced. Moreover, there is little change in the relative intensity of the 960 cm^{-1} band among MAS-6M, -7M and -8M glasses, which suggests that the amounts of MoO_4^{2-} in these glasses are quite close.

There are a number of sharp and split Raman peaks for the crystallised parts of phase separated glasses. These peaks are assigned to crystalline molybdates, and are different from the bands assigned to amorphous MoO_4^{2-} units in glass. Particularly, the patterns of crystalline peaks for CMAS-4M glass are almost identical to those for CAS-4M glass but distinct from those for MAS glass, indicating that CMAS-4M and CAS-4M glasses may have the same separated phase, most likely $CaMoO_4$. In contrast, the band assigned to amorphous MoO_4^{2-} units in CMAS-4M glassy part is centred at ~940 cm^{-1}, in the middle of the corresponding MoO_4^{2-} frequencies in MAS and CAS glasses which suggests that, when two alkaline earth species coexist in glass network, MoO_4^{2-} ions can be associated with both of them simultaneously and does not show strong preference to either one.

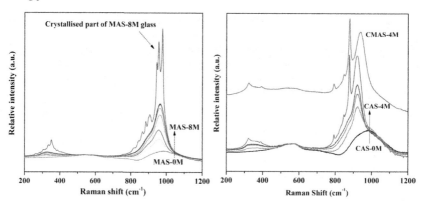

Figure 3 Normalised Raman spectra of (left) MAS glass and (right) CAS and CMAS glasses with different MoO_3 additions.

T_g Changes with MoO_3 incorporation

Glass transition temperature T_g is estimated from the onset of the first endothermic peak in DTA curves. As plotted in Figure 4, T_g is reduced from 775 °C for MAS base glass to 766 °C for MAS-1M glass and then continuously gradually reduced to 741 °C for MAS-8M glass which is phase separated. Similarly, T_g of CAS glass is reduced at initial MoO_3 addition, from 792 °C to 779 °C, and then gradually reduced to 771 °C for the partially phase separated CAS-4M glass. The reduced T_g suggests that the glass network is depolymerised by MoO_4^{2-} incorporation.

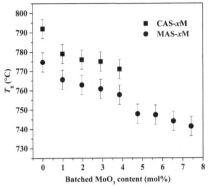

Figure 4 Changes in T_g of CAS and MAS glasses with increasing MoO_3 additions.

SEM

Figure 5 shows backscattered electron images of separated phases in MAS-8M and BAS-3M glasses. The separated particles are both spherical (droplet-like) and randomly dispersed within glass matrices, forming through liquid-liquid melt phase separation and thereafter crystallisation. The diameter of particles in MAS-8M glass ranges 100-300 nm while the diameter of particles in BAS-3M glass ranges 400-500 nm; both are smaller than the resolution limit (1 μm^2) of EDX measurement and thus compositions of these particles were not obtained.

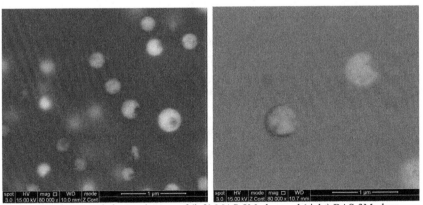

Figure 5 Backscattered electron images of (left) MAS-8M glass and (right) BAS-3M glass.

71

CONCLUSIONS

The MoO_3 solubility tendency in AeAS glasses suggests that the abundance of smaller alkaline earth helps to dissolve more MoO_3 without causing the occurrence of phase separation upon quenching. MAS glass shows the highest MoO_3 solubility of 5.34 mol% among all of them. MoO_3 retentions in AeAS glasses are notably high even though the glasses are melted at 1450 °C.

The incorporated MoO_4^{2-} ions in visibly homogeneous glasses and glassy parts of phase separated glasses remain amorphous according to XRD and Raman spectra. The Raman bands assigned to MoO_4^{2-} vibrations are intensified with increasing MoO_4^{2-} incorporation. The addition of MoO_3 also results in consistently reduced T_g, which means glass network is depolymerised by MoO_4^{2-} incorporation.

Excess MoO_3 addition leads to phase separation within glass matrices, forming randomly dispersed alkaline earth molybdate droplets hundreds of nanometres in diameter. Although MoO_4^{2-} ions show no preference in associating with Ca^{2+} or Mg^{2+} ions in the CMAS glass network, the separated phase in CMAS-4M glass seems to be predominantly $CaMoO_4$.

ACKNOWLEDGMENTS

ST thanks the "UK-China Scholarship for Excellence" for financial support. NCH is grateful to the Royal Academy of Engineering and Nuclear Decommissioning Authority for funding.

REFERENCES

1. B. F. Dunnett, N. R. Gribble, R. J. Short, E. Turner, C. J. Steele, A. D. Riley, *Glass Technol.: Eur. J. Glass Sci. Technol., Part A,* **54**, 166 (2012).
2. R. Do Quang, V. Petitjean, F. Hollebecque, O. Pinet, O. Flament, A. Prod'homme, WM'03 Conference, Tucson, AZ, 2003.
3. W. Lutze, R. C. Ewing, *Radioactive Wasteforms for the Future.* (North Holland, Amsterdam, 1988), p. 788.
4. M. I. Ojovan, W. E. Lee, *An Introduction to Nuclear Waste Immobilisation.* (Elsevier, Amsterdam, 2005), p. 250.
5. R. J. Short, R. J. Hand, N. C. Hyatt, G. Mobus, *J. Nucl. Mater.,* **340**, 179 (2005).
6. T. Taurines, B. Boizot, *J. Non-Cryst. Solids,* **357**, 2723 (2011).
7. C. M. Jantzen, *J. Non-Cryst. Solids,* **84**, 215 (1986).
8. I. Techer, T. Advocat, J. Lancelot, J. M. Liotard, *J. Nucl. Mater.,* **282**, 40 (2000).
9. J. E. Shelby, *J. Am. Ceram. Soc.,* **68**, 155 (1985).
10. M. Tiegel, A. Herrmann, C. Russel, J. Korner, D. Klopfel, J. Hein, M. C. Kaluza, *J. Mater. Chem. C,* **1**, 5031 (2013).
11. S. Tan, M.I. Ojovan, N.C. Hyatt, R.J. Hand, *J. Nucl. Mater.,* (2014) (in press).

Mater. Res. Soc. Symp. Proc. Vol. 1744 © 2015 Materials Research Society
DOI: 10.1557/opl.2015.299

Valence and Local Environment of Molybdenum in Aluminophosphate Glasses for Immobilization of High Level Waste from Uranium-Graphite Reactor Spent Nuclear Fuel Reprocessing

Sergey V. Stefanovsky[1], Andrey A, Shiryaev[1], Michael B. Remizov[2], Elena A. Belanova[2], Pavel A. Kozlov[2], and Boris F. Myasoedov[3]

[1] Frumkin Institute of Physical Chemistry and Electrochemistry RAS, Leninskii av. 31, Bld. 4, Moscow, 119071 Russia.
[2] FSUE Production Association "Mayak", Lenin st. 13, Ozersk Chelyabinsk reg. 456780 Russia
[3] Vernadsky Institute of Geochemistry and Analytical Chemistry RAS, Kosygin st. 19, Moscow 119071 Russia

ABSTRACT

Two Mo-bearing glasses considered as candidate forms for high level waste (HLW) a uranium-graphite reactor spent nuclear fuel (SNF) reprocessing were characterized. Incorporation of Mo in sodium aluminophosphate (SAP) glass increases its tendency to devitrification with segregation of orthophosphate phases. Valence state and local environment of Mo in the materials containing ~2 wt.% MoO_3 were determined by X-ray absorption fine structure (XAFS) spectroscopy. In the quenched samples composed of major vitreous and minor $AlPO_4$ nearly all Mo is located in the vitreous phase as $[Mo^{6+}O_6]$ units whereas in the annealed samples Mo is partitioned among vitreous and one or two orthophosphate crystalline phases in favor of the vitreous phase. Mo predominantly exists in a hexavalent state in distorted octahedral environment. Four oxygen ions are positioned at a distance of ~1.71-1.73 Å and two - at a distance of 2.02-2.04 Å. Minor Mo(V) is also present as indicated by a response in EPR spectra with $g \approx 1.911$-1.915.

INTRODUCTION

Currently in Russia some SNF compositions such as molybdenum-bearing SNF of uranium-graphite reactors (AMB) are not reprocessed yet but their reprocessing is under consideration now. HLW from AMB SNF reprocessing is suggested to be incorporated in SAP-based glass similarly to different current HLW [1]. Among the AMB fuels there are Mo-bearing varieties U+9%Mo/Mg, U+9%Mo/Ca, and U+3%Mo/Mg which will be reprocessed yielding Mo-containing HLW. Molybdenum is one of the troublesome components of HLW causing liquid/liquid phase separation in borosilicate glasses or devitrification of phosphate glasses thus reducing chemical durability of vitrified waste. Therefore, the effect of Mo solubility, its valence state and speciation on chemical durability of glasses has to be studied. Previous studies demonstrated a negative effect of MoO_3 on resistance to devitrification and chemical durability of SAP glasses [2].

There are numerous works on incorporation of Mo-bearing HLW in borosilicate glass (see, for example, [3-5]). Glass properties depend on Mo speciation in glass. Silicate and borosilicate glasses containing Mo(VI) have a very high tendency to liquid-liquid phase separation with formation of so-called "yellow phase" composed mainly of alkali and alkali earth molyddates, chromates, sulfates, and chlorides [6-8]. Maximum concentration of these anions is about 1 to 2 mol.% depending on glass composition.

Phosphate glasses have lower tendency to phase separation and various molybdate-, chromate-, sulfate-, and halogen-phosphate glasses are well-known [9-13]. Incorporation of alkali and alkali earth oxides in such glasses reduces glass forming areas and resistance of glasses to devitrification. Nevertheless, phosphate glasses remain to be promising matrices for wastes containing the elements of the VIth group of the Periodic Table in the highest oxidation states and halogenides.

There is a number of works on determination of molybdenum oxidation state and coordination in borosilicate glasses and their effect on Mo solubility in borosilicate [14-25] and phosphate [10-12,26-32] glasses for both nuclear and non-nuclear applications. In the present work, we study Mo speciation in complex SAP glasses considered as AMB HLW forms.

EXPERIMENTAL

Chemical composition of the glasses studied is given in Table I. The composition of the glass M0 approximately corresponds to that is currently produced in an EP-500 plant at the PA "Mayak"[1]. Nitrates of the elements were dissolved in a distilled water; the solutions were intermixed and dried in a dessicator. Dry mixtures were placed in silica crucibles, heated to 1000 °C in a resistive furnace, kept at this temperature for 0.5 hr, and then melts were poured onto a stainless steel plate (samples "q"). Pieces of the glasses were annealed at a temperature of 500 °C for 14 hrs (samples "a").

Table I. Target chemical composition of glasses (wt.%).

Glass ID	Na$_2$O	Cs$_2$O	P$_2$O$_5$	Al$_2$O$_3$	MoO$_3$	MgO	ZrO$_2$	Ce$_2$O$_3$	La$_2$O$_3$	Total
M0	20.4	3.9	54.2	21.3	0	0	0.07	0.03	0.1	100.0
M1	20.4	3.9	52.3	21.3	1.9	0	0.07	0.03	0.1	100,0
M2	18.6	3.9	50.5	19.5	1.9	5.4	0.07	0.03	0.1	100,0

The samples were characterized by X-ray diffraction using a Rigaku D/Max-2200 diffractometer (Cu Kα radiation), and scanning electron microscopy using a JSM-5610LV unit equipped with energy dispersive X-ray (EDX) spectrometer JED-2300.

Molybdenum speciation was determined by Mo K-edge X-ray absorption fine structure (XAFS) spectroscopy at the Structural Materials Science (STM) Beamline at the synchrotron source at NRC "Kurchatov Institute". The samples were measured at room temperature either as dispersed powders or as pellets compacted from powder mixed with sucrose in the transmission mode using a Si(220) channel-cut monochromator and two air-filled ionization chambers. Fluorescence spectra were also acquired. Powders of reagent-grade MoO$_2$ and MoO$_3$ were used as standards and measured under identical conditions. Experimental XAFS spectra were fitted in R-space using an IFEFFIT package [33] and crystal structures of corresponding crystalline compounds. In the fitting, *ab initio* photoelectron backscattering amplitudes and phases calculated self-consistently using FEFF8 [34] were used. Electron paramagnetic resonance (EPR) spectra were recorded at room temperature using a Bruker ESP-300 X-range spectrometer.

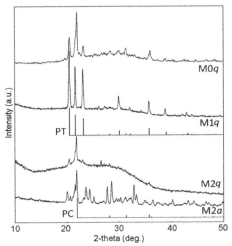

Figure 1. XRD patterns of the glassy samples. PC – phosphocrystobalite, PT – phosphotridymite. The rest of the reflections on M2*a* pattern are due to orthophosphate phases.

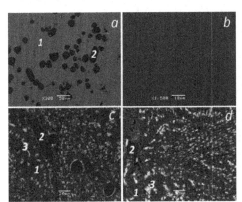

Figure 2. SEM images of the samples M1*q* (*a*), M2*q* (*b*), M1a (*c*), and M2*a* (*d*). 1 – glass (gray), 2 – AlPO$_4$ (black), 3 – High-Cs orthophosphates, Fine dendrite crystals are distributed in the vitreous phase.

RESULTS

XRD and SEM/EDX data

The quenched sample M1*q* is composed of an orthorhombic structure phase (phosphotridymite) with parameters close to pure AlPO$_4$ and vitreous phase. The quenched samples M0*q* M2*q* contained two aluminum orthophosphate varieties: major phosphocristobalite and minor phosphotridymite. All the annealed samples contained two or more orthophosphate crystalline phases (Fig. 1). The annealed sample M1a was predominantly crystalline while the sample M2a was composed of matrix vitreous phase (~40-45 vol.%) with primarily sodium aluminophosphate composition (gray on SEM images – Fig. 2), dendrite crystals of mixed Na-Mg aluminophosphate (~10-20 vol.%, light-gray), isometric and elongated nearly regular crystals of Al orthophosphates (phosphotridymite and phosphocrystobalite, respectively, 15-20 vol.% in total, dark-gray), and white grains (5-10 vol.%) enriched with Cs (Fig. 2). In the samples M1*q* and M2*q* all the Mo ions enter vitreous phase. In the samples M1a and M2a major Mo enters vitreous phase and minor one is partitioned among the two orthophosphate crystalline phases. Description of the XRD and SEM data in more details was given in our previous paper [36].

XAFS data

X-ray absorption near-edge structure (XANES) spectra (Fig. 3*a*) are very similar for all the samples studied exhibiting similar oxidation state and coordination environment of Mo ions. Comparison of the data obtained with reference data (Fig. 3*b* [22]) shows that the line-

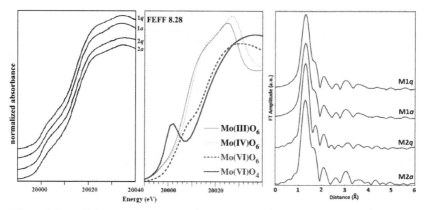

Figure. 3. XANES (*a* – our data, *b* – ref. data [22] and Fourier transform EXAFS (*c*) of Mo *K*-edge in glasses.

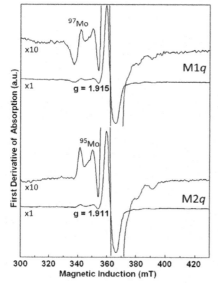

Figure 4. EPR spectra of the samples.

shape and major peak position, as well as absence of pre-edge peak typical of tetrahedrally coordinated Mo^{6+} ions, correspond to Mo^{6+} ions in octahedral oxygen environment.

Fourier transform Mo *K*-edge EXAFS spectra in all the samples studied are also similar (Fig. 3*c*). Mo^{6+} ions have an environment with a configuration of axially distorted octahedron with four shorter (1.71-1.73 Å) and two longer (2.02-2.04 Å) Mo–O distances. Only the first coordination shell is manifested pointing to homogeneous distribution of Mo^{6+} ions in glasses M1*q* and M2*q* and vitreous phases of glass-ceramics M1*a* and M2*a*. No contribution due to Mo in crystalline phases was observed.

EPR data

In the EPR spectra of the quenched samples (Fig. 4) the only single asymmetric line with g_{ef} = 1.915 (M1*q*) or g_{ef} = 1.915 (M2*q*) was present. This response may be attributed to Mo^{5+} ions in slightly distorted octahedral oxygen environment (molybdenyl-type complex). Minor contribution due to odd ^{95}Mo and ^{97}Mo isotopes (25% total abundance, I = 5/2) is also seen. Because glasses were produced in air concentration of Mo(V) is very low and no signals due to Mo(III) were detected.

DISCUSSION

The SAP glass has relatively high tendency to devitrification at elevated temperatures. If quenching of molten glasses with both M1 and M2 compositions yielded glassy products with low content of crystalline constituent (mainly $AlPO_4$) then annealing resulted in devitrification with segregation of normally two orthophosphate phases different in Cs_2O content. Major Mo enters vitreous phase where it is homogeneously distributed predominantly in the form of axially distorted $Mo^{6+}O_6$ octahedra. Similar Mo oxidation state and environment was revealed in different phosphate glasses [10,11,29] except work on lithium aluminophosphate glasses [32] where $Mo^{6+}O_4$ units were found while in the borosilicate glasses Mo(VI) preferably occupies tetrahedral sites [20-22,25]. In our glasses only minor Mo entered crystalline phases and its contribution cannot be evaluated. Thus possible reason of deterioration of chemical durability of SAP glasses at incorporation of MoO_3 observed in [1,2] is Mo concentrating in the residual vitreous phase.

EPR data show occurrence of minor Mo(V) which is present in the samples even produced in air. Mo^{5+} was firstly observed in borax glass [35] and later in silicate and borosilicate [14,16,17,19,22] and phosphate glasses [28,31] with similar EPR parameters. Mo(IV) not observable by EPR and Mo(III) may be also present in glasses produced under strongly reducing conditions as in [26,28]. Due to short time of spin-lattice relaxation the response from Mo(III) is normally observable at temperatures 77 K and lower.

CONCLUSIONS

Glasses containing AMB HLW surrogate and produced and quenched in air contain minor $AlPO_4$ phase but they are subject to devitrification under annealing with segregation of complex orthophosphate phases. Molybdenum in glassy materials enters predominantly residual vitreous phase and exists as Mo(VI) forming axially distorted MoO_6 octahedra with four equatorial oxygens at a distance of 1.71-1.73 Å and two oxygens at a distance of 2.02-2.04 Å. Minor Mo(V) is also present as it is seen from EPR response at $g \approx 1.911$-1.915. Concentrating of Mo in the residual vitreous phase of the devitrified glass may be a reason of deterioration of chemical durability of AMB HLW forms.

ACKNOWLEDGMENTS

The work was performed under financial support from Ministry of Science and Education of the Russian Federation under Agreement between the Ministry and Vernadsky Institute of Geochemistry and Analytical Chemistry No. 14.604.21.0009 dated 06/17/2014.

REFERENCES

1. E.A. Belanova, M.B. Remizov, A.S. Aloy, and T.I. Koltsova, *Problems Radiat. Safety* (Russ.) [2], 3 (2014).
2. E.A. Belanova, M.B. Remizov, A.S. Aloy, and T.I. Koltsova, *Problems Radiat. Safety* (Russ.) [4], 27 (2012).
3. O. Pinet, J.L. Dussossoy, C. David, and C. Fillet, *J. Nucl. Mater.* 377, 307 (2008).
4. B.F. Dunnett, N.R. Gribble, R. Short, E. Turner, C.J. Steele, and A.D. Riley, *Glass. Technol.: Eur. J. Glass Sci. Technol. A*, 53, 166 (2012).

5. R. Do Quang, V. Petitjean, F. Hollebecque, O. Pinet, T. Flament, and A. Prod'homme, in *Waste Management 2003 Conf.* February 23-27, 2003, Tucson, AZ (2003).
6. J.B. Morris, B.E. Chidley, in: *Management of Radioactive Wastes from the Nuclear Fuel Cycle.* Vienna, IAEA (1976).
7. W. Grunewald, H. Koschorke, S, Weisenburger, H. Zeh, in: *Radioactive Waste Management.* Proc. Int. Conf. Seattle, 16-20 May 1984. Vienna, IAEA, **2** (1984).
8. C. Gaudin, S. Schuller, L. Cormier, G. Calas, and S. Kroeker, in: *ATALANTE 2012.* Abstracts (2012), p. 301.
9. S.V. Stefanovsky, *Phys. Chem. Mater. Treat.* [2], 63 (1993).
10. G. Poirier, F.S. Ottoboni, F.C. Kassanjes, A. Remonte, Y. Messaddeq, and S.J.L. Ribeiro, *J. Phys. Chem.* **112**, 4481 (2008).
11. L. Koudelka, I. Rösslerová, J. Holubová, P. Mošner, L. Montagne, and B. Revel, *J. Non-Cryst. Solids,* **357**, 2816 (2011).
12. S. Marzouk, S.M. Abo-Naf, M. Hammam, Y.A. El-Gendy, and N.S. Hassan, *J. Appl. Sci. Res.* **7**, 935 (2011).
13. N. Da, O. Grassmé, K.H. Nielsen, G. Peters, L. Wondraczek, *J. Non-Cryst. Solids.* **357**, 2202 (2011).
14. B. Camara, W. Lutze, and J. Lux, "An Investigation on the Valency State of Molybdenum in Glasses with and without Fission Products," *Scientific Basis for Nuclear Waste Management,* ed. C.J.M. Northrup, Jr. (Plenum Press, 1980) **2**, pp. 93-102.
15. Y. Kawamoto, K. Clemens, and. M. Tomozawa, *J. Amer. Ceram. Soc.* **64**, 292 (1981).
16. Y. Kawamoto, K. Clemens, and. M. Tomozawa, and J.T. Warden, *Phys. Chem. Glasses* **22**, 110 (1981).
17. A. Horneber, B. Camara, and W. Lutze, *Mater. Res. Soc. Symp. Proc.* **11**, 279 (1982).
18. H.D. Schreiber, *J. Geophys. Res.* **92**, 9225 (1993).
19. R.J. Short, R.J. Hand, and N.C. Hyatt, *Mater. Res. Soc. Symp. Proc.* **757**, 141 (2002).
20. G. Calas, M. Le Grand, L. Galoisy, and D. Ghaleb, *J. Nucl. Mater.* **322**, 15 (2003).
21. R.J. Short, R.J. Hand, N.C. Hyatt, and G. Möbus, *J. Nucl. Mater.* **340**, 179 (2005).
22. F. Farges, R. Siewert, G.E. Brown, Jr., A. Guesdon, and G. Morin, *Canad. Miner.* **44**, 731 (2006).
23. F. Farges, R. Siewert, C.W. Ponader, G.E. Brown, Jr., M. Pichavant, and H. Behrens, *Canad. Miner.* **44**, 755 (2006).
24. S. Schuller, O. Pinet, A.Granjiean, and T. Blisson, *J. Non-Cryst. Solids* **354**, 296 (2008).
25. D. Caurant, O. Majérus, E. Fadel, A. Quintas, C. Gervais, T. Charpentier, and D. Neuville, *J. Nucl. Mater.* **396**, 94 (2010).
26. R.J. Landry, *J. Chem. Phys.* **48**, 1422 (1968).
27. S. Parke and A.C. Watson, *Phys. Chem. Glasses* **10**, 37 (1969).
28. J. Baucher and S. Parke, "ESR and Optical Studies of Mo(V) in Phosphate Glasses," *Amorphous Materials,* ed. R.W. Douglas and B. Ellis (Wiley, 1971) pp. 399-404.
29. A. Kuzmin and J. Purans, *J. Phys. IV France* **7**, C2-971 (1997).
30. G.D. Khattak, M.A. Salim, A.S. Al-Harthi, D.L. Thompson, and L.E. Wenger, *J. Non-Cryst. Solids* **212**, 180 (1997).
31. O. Cozar, D.A. Magdas, and I. Ardelean, *J. Non-Cryst. Solids* **354**, 1032 (2008).
32. Y.B. Saddeek and S.M. Abo-Naf, *Archives of Acoustics,* 37, 341 (2012).
33. B. Ravel and M. Newville, *J. Synchrotron Radiat.* **12** 537-541 (2005).
34. A.L. Ankudinov and J.J. Rehr, *Phys. Rev. B* **56** 1712-1716 (1997).
35. N.S. Garif'yanov and N.R. Yafaev, *Sov. Phys. – JETF* **16**, 1392 (1963).

Mater. Res. Soc. Symp. Proc. Vol. 1744 © 2015 Materials Research Society
DOI: 10.1557/opl.2015.312

Copper Valence and Local Environment in Aluminophosphate Glass-Ceramics for Immobilization of High Level Waste from Uranium-Graphite Reactor Spent Nuclear Fuel Reprocessing

Sergey V. Stefanovsky[1], Andrey A, Shiryaev[1], Michael B. Remizov[2], Elena A. Belanova[2], Pavel A. Kozlov[2], Boris F. Myasoedov[3]
[1] Frumkin Institute of Physical Chemistry and Electrochemistry RAS, Leninskii av. 31, Bld. 4, Moscow, 119071 Russia.
[2] FSUE Production Association "Mayak",
Lenin st. 13, Ozersk Chelyabinsk reg. 456780 Russia
[3] Vernadsky Institute of Geochemistry and Analytical Chemistry RAS, Kosygin st. 19, Moscow

ABSTRACT

Copper valence and environment in two sodium aluminophosphate glasses suggested for immobilization of HLW from reprocessing of spent fuel of uranium-graphite channel reactor (Russian AMB) were studied by XRD, SEM/EDX, XAFS and EPR. Target glass formulations contained ~2.4-2.5 mol.% CuO. The quenched samples were predominantly amorphous. The annealed MgO free sample had higher degree of crystallinity than the annealed MgO-bearing sample but both them contained orthophosphate phases. Cu in the materials was partitioned in favor of the vitreous phase. In all the samples copper is present as major Cu^{2+} and minor Cu^+ ions. Cu^{2+} ions form planar square complexes (CN=4) with a Cu^{2+}-O distance of 1.93-1.95 Å. Two more ions are positioned at a distance of 2.76-2.86 Å from Cu^{2+} ions. So the Cu^{2+} environment looks like a strongly elongated octahedron as it also follows from the absence of the pre-edge peak due to $1s{\rightarrow}3d$ transition in Cu K edge XANES spectra of the materials. Cu^+ ions form two collinear bonds at Cu^+-O distances of 1.80-1.85 Å. Thus average Cu coordination number (CN) in the first shell was found to be 2.7-3.0.

INTRODUCTION

High-level waste (HLW) from reprocessing of spent nuclear fuel of uranium-graphite reactors (Russian AMB type) is suggested to be vitrified with production of sodium aluminophosphate based glass similarly to current PWR (Russian WWER) waste [1]. One of the AMB fuel compositions is a UO_2+Cu/Mg (urania fuel with Cu/Mg subbed) and therefore behavior of copper ions in sodium aluminophosphate glasses has to be investigated. Copper speciation in phosphate glasses was studied earlier (see, for example [2-14]) but only few works [6,7,9,10] were concerned to alkali phosphate glasses. In glasses prepared under oxidizing conditions (in air) major Cu is present as Cu^{2+} ions and capable to form P–O–Cu bridging bonds which can replace P–O···Na^+ bonds enhancing the chemical durability of the glasses. EPR studies have demonstrated occurrence of Cu^{2+} ions in both silicate and phosphate glasses (see, for example, [15,16] and review [17]). The most important two conclusions were made: Cu^{2+} ion has a tendency to dictate its own environment in many oxide glasses [17] and Cu^{2+} ions are partitioned among two co-existing phases in phase-separated glasses occupying two unequivalent sites and produce two EPR spectra with essentially different parameters [16].

In the present work we studied complex sodium aluminophosphate glasses containing HLW surrogate from AMB spent fuel reprocessing. The goal of the work is to determine Cu speciation in glasses and glass crystalline materials formed at vitrification of HLW surrogates. Some preliminary results of a study of the phase composition and elemental partitioning among coexisting phases in these materials were described in refs [18-20].

EXPERIMENTAL

Chemical composition of the glasses studied is given in Table I. The baseline glass composition (approximately, in wt.%, 24 Na_2O, 23 Al_2O_3, 53 P_2O_5) is used at PA "Mayak" as a matrix for HLW immobilization [1]. Nitrates of the elements were dissolved in a distilled water; the solutions were intermixed and dried in a dessicator. Dry mixtures were placed in silica crucibles, heated to 1000 °C in a resistive furnace, kept at this temperature for 0.5 hr, and then melts were poured onto a stainless steel plate (samples "q"). Pieces of the glasses were annealed at a temperature of 500 °C for 14 hrs (samples "a").

Table I. Target chemical composition of glasses (wt.%).

Glass ID	Na_2O	Cs_2O	P_2O_5	Al_2O_3	CuO	MgO	ZrO_2	Ce_2O_3	La_2O_3	Total
1	20.4	3.9	52.3	21.3	1.9	0	0.07	0.03	0.1	100,0
2	18.6	3.9	50.5	19.5	1.9	5.4	0.07	0.03	0.1	100,0

The samples were characterized by X-ray diffraction using a Rigaku D/Max-2200 diffractometer (Cu Kα radiation), and scanning electron microscopy using a JSM-5610LV unit equipped with energy dispersive X-ray (EDX) spectrometer JED-2300.

Copper speciation was determined by X-ray absorption fine structure (XAFS) spectroscopy at the Structural Materials Science (STM) Beamline at the synchrotron source at NRC "Kurchatov Institute". The glass samples were measured at room temperature either as dispersed powder or as pellets compacted from powder mixed with sucrose in the transmission mode using a Si(220) channel-cut monochromator and two air-filled ionization chambers. Fluorescence spectra were also acquired. Powders of reagent grade Cu_2O, CuO and $CuCl_2·2H_2O$ were used as standards and measured under identical conditions. Experimental XAFS spectra were fitted in R-space using an IFEFFIT package [21] and crystal structures of corresponding oxides. In the fitting, *ab initio* photoelectron backscattering amplitudes and phases calculated self-consistently using FEFF8 [22] were used. Electron paramagnetic resonance (EPR) spectra were recorded at room temperature using a Bruker ESP-300 X-range spectrometer.

RESULTS

XRD and SEM/EDX data

XRD patterns exhibit nearly amorphous state of the quenched samples and some devitrification of the annealed samples (Fig. 1). The sample $1q$ contains minor (20-30 vol.% from SEM data) $AlPO_4$ in the form of phosphotridymite. On XRD pattern of the sample $2q$ there is the only reflection due to phosphocrystobalite. As follows from XRD (Fig. 1) and SEM/EDX (Fig. 2) data, the sample $1a$ is composed of phosphotridymite (the darkest crystals on SEM image), Na/Al orthophosphate (gray crystals), and Na/Cs/Al orthophosphate (white grains) distributed in the vitreous phase. Chemical composition of these phases is recalculated well to formulae $Al_{0.97}P_{1.03}O_{4.03}$, $Na_{0.75}Cs_{0.02}Cu_{0.02}Al_{0.60}P_{1.07}O_{3.98}$ and $Na_{0.72}Cs_{0.07}Cu_{0.04}Al_{0.54}P_{1.10}O_{4.00}$, respectively. The vitreous phase is enriched with Cs and Cu relatively to orthophosphates. The sample $2a$ has similar phase composition Phosphocrystobalite (black on SEM image) and two different Na/Cs//Mg/Al orthophosphate (dark-gray and light on SEM image) crystals are distributed in the vitreous phase (light-gray on SEM image). Formulae of these phases are as follows: $Al_{0.93}Mg_{0.05}Na_{0.06}P_{1.01}O_{4.00}$, $Na_{0.61}Mg_{0.13}Cs_{0.11}Cu_{0.02}Al_{0.53}P_{1.08}O_{4.00}$, and $Na_{0.63}Cs_{0.24}Mg_{0.21}Cu_{0.02}Al_{0.52}P_{1.02}O_4$, respectively. XRD pattern for one of the mixed phosphate phase is similar to that of $Na_3Al_2(PO_4)_3$ (or $NaAl_{0.67}PO_4$), for other phase no reference XRD pattern has been found. Thus, in the quenched samples all the copper ions enter the vitreous phase while in the annealed samples Cu is partitioned among vitreous and crystalline phases in favor of the first of them especially in the sample $1a$ with very low content of Cu-bearing Na/Cs/Al orthophosphate phase.

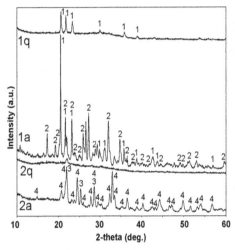

Figure 1. XRD patterns of the samples.
1 – phosphotridymite, 2 – Na/Al orthophosphate of $Na_3Al_2(PO_4)_3$ type, 3 – phosphocrystobalite, 4 – complex Na/Cs/Mg/Al-phosphate.

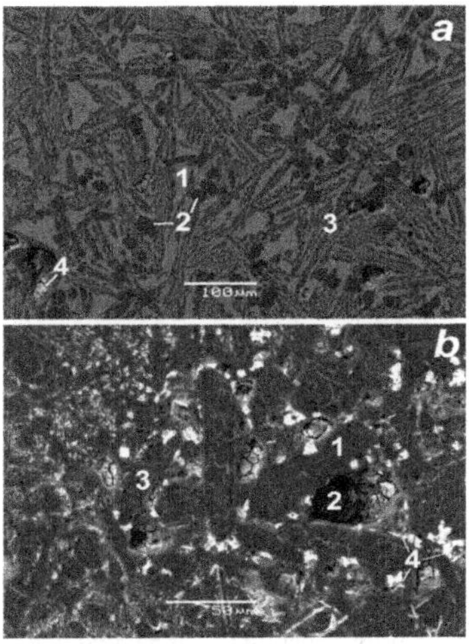

Figure 2. SEM images of the samples $1a$ (a) and $2a$ (b).
1 – glass (gray), 2 – $AlPO_4$ (black), 3 - Na/Al orthophosphate, 4 - complex Na/Cs/Mg/ Al-phosphate

XAFS data

X-ray absorption near-edge structure (XANES) spectra (Fig. 3) are very similar for all the samples studied and typical of Cu^{2+} ions. The contribution due to Cu^+ ions is rather low ($\leq 20\%$). Because no difference between the quenched amorphous and annealed partly devitrified samples it may be concluded that major copper ions enter the vitreous phase and occupy similar structural sites. Absence of the pre-edge shoulder due to the $1s \rightarrow 3d$ transition points to distorted octahedral oxygen environment of Cu^{2+} ions [23] forming planar square complexes, where four oxygens are positioned at a distance of 1.93-1.95 Å from central Cu^{2+} ion and two more oxygens are positioned at a distance of 2.76-2.86 Å from Cu^{2+} ion with total CN=6 (Fig. 4 and Table II).

Cu$^+$ ions form two collinear bonds at Cu$^+$–O distances of 1.80-1.85 Å. Thus average Cu coordination number (CN) in the first shell was found to be 2.7-3.0.

Unlike crystalline oxides, no second coordination shell appeared in the FT EXAFS spectra of both the quenched and annealed samples pointing to predominant incorporation of Cu in the vitreous phase and homogeneous distribution of copper ions in its structure.

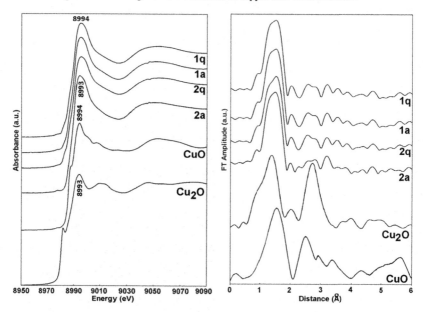

Figure 3. XANES spectra of Cu K-edge in the samples studied and standards.

Figure 4. Fourier transforms EXAFS of Cu K-edge in the samples studied and standards.

Table II. Computer simulation of the first coordination shell of copper ions in the aluminophosphate glassy materials.

Glass ID	R_{Cu-O}, Å	CN	Glass ID	R_{Cu-O}, Å	CN
1q	1.94	2.78	2q	1.93	2.73
	2.76	0.62		2.78	0.7
1a	1.95	2.86	2a	1.93	2.98
	2.79	0.81		2.86	0.74

EPR data

EPR spectra consist of two sets each of four lines caused by hyperfine splitting due to ^{63}Cu and ^{65}Cu (I=3/2 for both isotopes) (Fig. 5). This is a typical response due to Cu^{2+} ions ($3d^9$

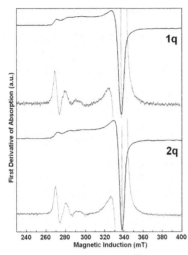

Figure 5. EPR spectra of Cu^{2+} ions in glasses.

electronic configuration, ^2D ground state) in a strongly distorted octahedral environment with a geometrical configuration of tetragonally elongated octahedron [17]. The spin-Hamiltonian parameters obtained from room temperature spectra are $A_\parallel = 12.5$ mT, $g_\parallel = 2.383$, $g_\perp = 2.079$ (for the sample 1q) and $A_\parallel = 12.3$ mT, $g_\parallel = 2.384$, $g_\perp = 2.074$ (for the sample 2q) and seem to be very similar exhibiting similar local environment of the Cu^{2+} ions in glass network.

DISCUSSION

Copper-bearing sodium aluminophosphate glass has elevated tendency to devitrification but this depends mainly on glass composition, which is located between orthophosphate and pyrophosphate compositional lines in the Na_2O-Al_2O_3-P_2O_5 system. Minor MgO substitution for a sum of $Na_2O+Al_2O_3+P_2O_5$ increases resistance of glass to devitrification but does not influence on Cu partitioning because predominant Cu enters the vitreous phase. We have not found splitting of the EPR responses due to partitioning of Cu^{2+} ions between co-existing phases.

In general Cu behavior in sodium aluminophosphate glasses both MgO free and MgO-bearing is typical of that in different phosphate glasses [12,13,16,17]. In our glasses and glass-ceramics similarly to different glasses major Cu exists as Cu^{2+} ions in the strongly distorted octahedral environment while minor Cu is probably present as linear complexes O–Cu–O. As follows from FT EXAFS data, copper ions are homogeneously distributed in glass and in the vitreous phases of the glass-ceramics. This was also confirmed by EPR data because no dipole-dipole interaction between Cu^{2+} appearing as a line broadening was revealed.

Due to low concentration of copper oxides in the materials, we were unable to evaluate their effect on the structure of the anionic motif of the glass network and liquid-liquid phase separation and devitrification of the glasses or their chemical durability. Anyway minor copper enters crystalline phases as isomorphic dopant but does not form own copper-bearing phases.

CONCLUSIONS

Copper-bearing sodium aluminophosphate glass with baseline composition has rather high tendency to devitrification. Incorporation of 5.4 wt.% MgO instead of all other components makes glass more resistant to devitrification. In the glass-ceramics major Cu enters vitreous phase as Cu^{2+} ions in the strongly distorted octahedral environment where four equatorial oxygen ions are positioned at a distance of 1.93-1.95 Å and two more axial oxygens – at a distance of 2.76-2.86 Å from Cu^{2+} ion. Minor Cu exists as Cu^+ ions as linear complexes with Cu^+–O distances of 1.80-1.85 Å. Cu ions are homogeneously distributed in the vitreous phase not forming clusters.

ACKNOWLEDGMENTS

The work was performed under financial support from Ministry of Science and Education of the Russian Federation (RFMEFI60414X0009).

REFERENCES

1. S.V. Stefanovsky, S.V. Yudintsev, R. Giere, G.R. Lumpkin, "Nuclear Waste Forms," *Energy, Waste and the Environment: A Geological Perspective*, ed. R. Giere and P. Stille (Geological Society, London, 2004) pp. 37-63.
2. A. Musinu, G. Piccaluga, G. Pinna, G. Vlaic, D. Narducci, and S. Pizzini, *J. Non-Cryst. Solids*, **136**, 198 (1991).
3. B.-S. Bae and M.C. Weinberg, *J. Non-Cryst. Solids*, **168**, 223 (1994).
4. G.D. Khattak, M.A. Salim, A.B. Hallak, M.A. Daous, L.E. Wenger, and D.J. Thompson, *J. Mater. Sci.* **30**, 4032 (1995).
5. J. Koo, B.-S. Bae, and H.-K. Na, *J. Non-Cryst. Solids*. **212**, 173 (1997).
6. P.Y. Shih, S.W. Yung, and T.S. Chin, *J. Non-Cryst. Solids*. **224**, 143 (1998).
7. P.Y. Shih, S.W. Yung, and T.S. Chin, *J. Non-Cryst. Solids*. **244**, 211 (1999).
8. E. Metwalli, M. Karabulut, D.L. Sidebottom, M.M. Morsi, and R.K. Brow, *J. Non-Cryst. Solids*. **344**, 128 (2004).
9. A. Chahine, M. El-Tabirou, and J.L. Pascal, *Mater. Lett.* **58**, 2776 (2004).
10. A. Chahine, M. El-Tabirou, M. Elbenaissi, M. Haddad, and J.L. Pascal, *Mater. Chem. Phys.* **84**, 341 (2004)
11. U. Hoppe, R. Kranold, A. Barz, D. Stachel, A. Schöps, and A.C. Hannon, *Phys. Chem. Glasses.* **48**, 188 (2007).
12. G. Giridhar, M. Ragnacharyulu, R.V.S.S.N. Ravikumar, and P. Sambasiva Rao, *J. Mater. Sci. Technol.* **25**, 531 (2009).
13. N. Vedeanu, D.A. Magdas, and R. Stefan, *J. Non-Cryst. Solids*. **358**, 3170 (2012).
14. D.A. Magdas, R. Stefan, D. Toloman, and N. Vedeanu, *J. Mol. Struct.* **1056-1057**, 314 (2014).
15. L.D. Bogomolova, V.F. Jachkin, V.N. Lazukin, and V.A. Schmukler, *J. Non-Cryst. Solids*. **27**, 427 (1978).
16. L.D. Bogomolova, V.F. Jachkin, V.N. Lazukin, T.K. Pavlushkina, and V.A. Schmukler, *J. Non-Cryst. Solids*. **28**, 375 (1978).
17. D.L. Griscom, *J. Non-Cryst. Solids*. **40**, 211 (1980).
18. M.B. Remizov, E.A. Belanova, S.V. Stefanovsky, B.F. Myasoedov, B.S. Nikonov, *Glass Phys. Chem.* **40**, 534 (2014).
19. S.V. Stefanovsky, B.S. Nikonov, M.B. Remizov, P.V. Kozlov, E.A. Belanova, A.A. Shiryaev, and Ya.V. Zubavichus, *Phys. Chem. Mater. Treat.* (Russ.) [5], 74 (2014).
20. S.V. Stefanovsky, B.F. Myasoedov, M.B. Remizov, P.V. Kozlov, E.A. Belanova, A.A. Shiryaev, and Ya.V. Zubavichus, *Doklady Chemistry*, **457**, 148 (2014).
21. B. Ravel and M. Newville, *J. Synchrotron Radiat.* **12**, 537-541 (2005).
22. A.L. Ankudinov and J.J. Rehr, *Phys. Rev. B* **56** 1712-1716 (1997).
23. L.A. Grunes, *Phys. Rev.* **B27**, 2111-2131 (1983).

Mater. Res. Soc. Symp. Proc. Vol. 1744 © 2015 Materials Research Society
DOI: 10.1557/opl.2015.382

Nepheline Crystallization in High-Alumina High-Level Waste Glass

José Marcial[1], John Mccloy[1], Owen Neill[2]
[1]School of Mechanical & Materials Engineering & Materials Science and Engineering Program,
Washington State University, Pullman, WA 99164-2920, USA

[2]Peter Hooper GeoAnalytical Laboratory, School of the Environment
Washington State University
Pullman, WA 99164-2812, USA

ABSTRACT

The understanding of the crystallization of aluminosilicate phases in nuclear waste glasses is a major challenge for nuclear waste vitrification. Robust studies on the compositional dependence of nepheline formation have focused on large compositional spaces with hundreds of glass compositions. However, there are clear benefits to obtaining complete descriptions of the conditions under which crystallization occurs for specific glasses, adding to the understanding of nucleation and growth kinetics and interfacial conditions. The focus of this work was the investigation of the microstructure and composition of one simulant high-level nuclear waste glass crystallized under isothermal and continuous cooling schedules. It was observed that conditions of low undercooling, nepheline was the most abundant aluminosilicate phase. Further undercooling led to the formation of additional phases such as calcium phosphate. Nepheline composition was independent of thermal history.

INTRODUCTION

In typical nuclear waste vitrification, multiphase solid/liquid waste feed is mixed with glass-forming additives (e.g., SiO_2 and H_3BO_3) and converted to glass for storage in a permanent geological repository. After feed-to-glass conversion, molten glass is poured into stainless steel canisters and air quenched [1]. The cooling profile of the canister allows for rapid cooling of the outer melt but results in slow cooling of melt in the canister center [1]. Often this behavior is simulated with benchtop-scale testing by heat treating glasses following a canister-centerline cooling (CCC) schedule. As dictated by Johnson-Mehl-Avrami-Kolmogarov kinetics, crystallization of aluminosilicate phases will proceed given sufficient time and temperature for lattice arrangement of $[SiO_4]^{4-}$ and $[AlO_4]^{5-}$ tetrahedra [1]. The aluminosilicate phase of interest is nepheline, $NaAlSiO_4$, which has been previously shown to be deleterious to chemical durability due to the extraction of alumina and silica from the glass-forming matrix, leaving a residual glass of less-durable components [2-5]. Nepheline can accept up to 0.25 atomic% K as $Na_{0.75}K_{0.25}AlSiO_4$. The long-term corrosion resistance is significant because vitrified waste must tolerate subterranean storage conditions for $\geq 10^6$ years with limited radioisotope transport.

To understand the crystallization behavior of glass, isothermal heat treatments (IHT) were performed at 1050-750°C for 3-30 hours. Compositional analysis of crystallized glasses was performed through wavelength-dispersive spectroscopy (WDS). Phase identification was performed through X-ray diffraction (XRD). Scanning electron microscopy energy-dispersive spectroscopy (SEM-EDS) and SEM-WDS maps were further collected to examine the segregation behavior of major elements.

METHODS

In this study a non-radioactive simulant high-level waste glass designated "A4" was selected for investigation after observation of its unique microstructure upon crystallization [6]. Table I displays the batched glass composition and table II provides the measured glass composition. A4 was formulated for a high-alumina waste stream (>25 wt% Al_2O_3) with 45 wt% waste loading, which is typically a very challenging composition for glass formation. Samples were batched from powder precursors and melted at 1200°C in a Pt-10% Rh crucible for 1 hour followed by air quenching, IHT, or CCC to observe the crystallization behavior. Both IHT and CCC were performed using 15g of glass frit in 1×1×1" Pt-10% Rh boxes. The IHT were performed at 1050, 900, 825, and 750°C for dwell times ranging 3-30 hours (Figure 1). The 825°C-30 hour sample only exhibited evidence of nepheline crystallization on the meniscus and at the glass interface with the platinum crucible (i.e., heterogeneous nucleation). Figure 1 presents a schematic of the sample geometry after CCC heat treatment [1]. XRD was performed with a PANalytical X'Pert PRO MPD X-ray Diffractometer outfitted with a Co-Kα X-ray source and incident beam monochromator operated at 45 KV and 40 mA. The scan parameters were: 20-100° 2θ, 0.005° step, and 20 s dwell. For the 750 and 825°C IHT samples, the nephelines identified were hexagonal Si-rich nephelines (ICDD 98-010-8334 for 750°C and 01-075-2934 for both 750 and 825°C). Two nephelines were identified in the CCC sample, one was a rhombohedral nepheline bearing K and Ca (ICDD 05-001-0051) and the other was Si-rich, hexagonal, and K-bearing (ICDD 98-003-7356).

Table I. Batched A4 glass composition   Table II. WDS measured A4 glass composition

Constituent	Mass Fraction	Constituent	Mass Fraction	Constituent	Mass Fraction	Constituent	Mass Fraction
SiO_2	0.305	Cr_2O_3	0.005	SiO_2	0.302±0.006	Cr_2O_3	0.000±0.000
Al_2O_3	0.240	PbO	0.004	Al_2O_3	0.262±0.003	PbO	0.004±0.000
B_2O_3	0.120	NiO	0.004	B_2O_3	0.155±0.001	NiO	0.000±0.000
Na_2O	0.096	ZrO_2	0.004	Na_2O	0.083±0.006	ZrO_2	0.004±0.000
Li_2O	0.068	SO_3	0.002	Li_2O	0.069	SO_3	0.002±0.000
CaO	0.061	K_2O	0.001	CaO	0.064±0.001	K_2O	0.001±0.000
Fe_2O_3	0.059	MgO	0.001	Fe_2O_3	0.028±0.001	MgO	0.001±0.000
Bi_2O_3	0.011	ZnO	0.001	Bi_2O_3	0.011±0.000	ZnO	0.000±0.000
P_2O_5	0.011	Total	1.000	P_2O_5	0.011±0.000	Total	1.000
F	0.007			F	0.002±0.001		

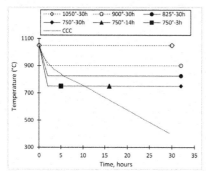 

Figure 1 (left). Graphical depiction of IHT and CCC heat treatment schedules of A4. The darkened IHT points represent the samples which exhibited observable crystallization. **Right** Optical image of A4-CCC.

For SEM-EDS and SEM-WDS, all samples were sectioned and polished with 400, 800, and 1200 grit SiC paper, then with 3 and 1 μm diamond paste. Samples were analyzed on a JEOL JXA-8500F field-emission electron microprobe/scanning electron microscope. Beam conditions for measurement of nepheline were 15 KeV accelerating voltage, 8 nA probe current and 7 μm spot size. Measurements were performed on the largest observed nepheline branches. Table III provides the experimental parameters for the WDS measurements of nepheline. For measurements of A4 glass the beam conditions were 20 KeV accelerating voltage, 10 nA probe current and 10 μm spot size. Table IV provides the experimental parameters for the measurement of A4 glass. For the measurements of glass, the concentration of Li was calculated by difference.

Heat treated samples were imaged through SEM and analyzed through WDS for the concentration of Na, K, Ca, Mg, Fe, B, Al, Si, and Zr in nepheline. Figure 2 provides microstructural images of A4-CCC, A4-825-30 h, and A4-750-30 h. Maps collected of A4-CCC through SEM-EDS and SEM-WDS are shown in Figure 3. Zirconium data was collected to discriminate between glass and nepheline after determining through SEM-EDS that nepheline did not contain Zr.

Table III. Spectrometer conditions and standard assignments for WDS measurements of A4 nepheline

Element/X-ray Line	Analyzing Crystal	Peak Count Time, s	Background Count Time, s	Standard
Na Kα	TAP	20	10	Albite #4 (C.M. Taylor)
Si Kα	TAP	20	10	NIST Glass K-412
Al Kα	TAP	20	10	Spessartine (C.M. Taylor)
Mg Kα	TAP	150	75	Diopside #1 (C.M. Taylor)
Fe Kα	LiF	240	120	Hornblende, Wilburforce
K Kα	PETH	60	30	Hornblende, Kakanui
Zr Lα	PETH	120	60	Zircon #1 (C.M. Taylor)
Ca Kα	PETJ	240	120	Diopside #1 (C.M Taylor)
B Kα	LDE6 (Cr/C Synthetic Multilayer, 2d = 120 Å)	240	120	NIST Glass K-490

Table IV. Spectrometer conditions and standard assignments for WDS measurements of A4 glass

Element/X-ray Line	Analyzing Crystal	Peak Count Time, s	Background Count Time, s	Standard
Na Kα	TAP	15	7.5	NIST Glass K-373
Al Kα	TAP	15	7.5	Chromite #5, Australia
Si Kα	PETJ	15	7.5	NIST Glass K-373
Fe Kα	LiF	20	10	Pentlandite (Astimex)
Zr Lα	PETH	20	10	Zircon #1 (C.M. Taylor)
Ca Kα	PETJ	20	10	Apatite, Wilburforce
B Kα	LDE6 (Cr/C Synthetic Multilayer, 2d = 120 Å)	240	120	NIST Glass K-490
K Kα	PETH	30	15	Orthoclase (MAD-10)
Mg Kα	TAP	20	10	MgF$_2$ (C.M. Taylor)
Ni Kα	LiF	20	10	Pentlandite (Astimex)
Zn Kα	LiF	20	10	Willimite #1 (C.M. Taylor)
Cr Kα	LiF	20	10	Chromite #5, Australia
Pb Mα	PETH	20	10	PbS (C.M. Taylor)
S Kα	PETH	20	10	Pentlandite (Astimex)
P Kα	PETJ	20	10	Apatite, Wilburforce
Bi Mα	PETJ	20	10	Bi$_{12}$GeO$_{20}$ (C.M. Taylor)
F Kα	TAP	20	10	MgF$_2$ (C.M. Taylor)

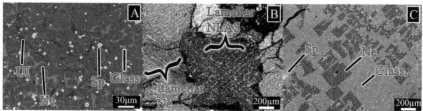

Figure 2. Backscatter electron (BSE) microstructural images of heat treated A4 samples: (A) CCC sample, (B) 750°C-30 hour and (C) 825°C-30 hour sample where: UI=unidentified phase, Ne=nepheline, NLAS=nepheline-like aluminosilicate, and Sp=spinel. Note the smaller scale for left image

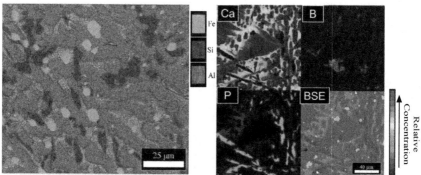

Figure 3. (Left) SEM EDS map of A4 CCC, (Right) SEM WDS maps of A4 CCC; BSE indicates backscattered electron image

DISCUSSION

Backscatter electron (BSE) images of the microstructural features of the three samples in Figure 2 provide insight into the phases present in each sample. The 825°C-30 h sample, which experienced a lower degree of undercooling, only exhibited these three phases; however, the CCC and the 750°C-30 h samples featured additional phases which were difficult to characterize due to small crystallite size, low volume fraction, and challenges associated with the presence of light elements.

Compositional mapping of the CCC sample revealed the presence of phosphorus-rich, boron-rich, and spinel phases between nepheline lamellae. Regions that exhibit boron enrichment also exhibit a different contrast. The presence of calcium-phosphorus regions was also observed in the 750°C-30 h sample as determined through SEM-EDS analysis is shown in Figure 4. Furthermore, the 750°C-30 h sample featured an additional nepheline-like aluminosilicate (NLAS) phase which could not be identified through SEM-WDS (only Na, Al, Si, and O were detected). However, based on the lithium content of this system, it is likely that this is a lithium aluminosilicate phase. Li is not detectable by EDS or WDS. Further justification for this assumption can be found in the modelling of the nepheline-spinel psuedobinary phase diagram for high-level waste borosilicate glasses by Jantzen and Brown [7]. From the work of Jantzen and Brown, hypoeutectic compositions were found to readily form various primary and secondary silicate phases, some of which are Li-bearing [7]. Anomalously, there exist P-enriched regions in the CCC sample that are devoid of calcium, unlike the Calcium phosphate phase in Figure 4.

The relevance of the existence of the additional crystalline phases to the durability of the glass requires further investigation. Certain crystalline phases, namely spinels, are not detrimental to glass durability; however, NLAS are believed to compromise chemical durability to a similar extent as nepheline [2-3].

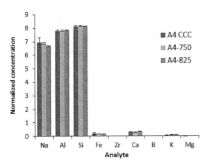

	Ca-P	Spinel	NLAS
	Element Wt%		
O K	39.59	23.32	50.48
Na K	0.05±0.02	0.34±0.04	
Mg K		0.30±0.03	
Al K		4.82±0.04	23.77±0.18
Si K	0.68±0.02	0.39±0.02	25.75±0.21
P K	16.79±0.13		
Fe K		42.97±0.36	
Ca K	42.88±0.21		
Cr K		15.01±0.17	
Ni K		12.85±0.23	
Total	99.99	100.0	100.0

Figure 4. BSE and semi-quantitative SEM-EDS of A4-750°C-30 h exhibiting multiple crystalline phases (Sp=spinel, Ca-P=calcium phosphate, Ne=nepheline; NLAS=nepheline-like aluminosilicate)

Figure 5 compares the compositions of nepheline found in the CCC, 750°C-30h, and 825°C-30h samples as obtained through SEM-WDS and normalized to 32 oxygen atoms. Despite the significant differences in the microstructure of these samples (Figure 2), the compositional variation is less drastic. This is perhaps indicative of the normative behavior of this melt system. If we first assume that the conditions of the 825°C-30 h sample are within the undercooling-controlled regime, then the presence of nepheline as the only aluminosilicate phase is indicative of a thermodynamic preference for the nepheline nucleation. The similarity in composition of the crystallites may then suggest that the subsequent crystallization of other aluminosilicate phases proceeds by the local enrichment of solute atoms.

	A4 CCC	A4-750	A4-825
	Average:	Average:	Average:
Na	6.94±0.35	6.89±0.06	6.66±0.06
Al	7.79±0.09	7.81±0.05	7.84±0.03
Si	8.12±0.06	8.14±0.03	8.14±0.03
Fe	0.19±0.08	0.16±0.02	0.17±0.01
Ca	0.31±0.02	0.28+0.02	0.34±0.01
B	0.00	0.00	0.00
K	0.06±0.00	0.09±0.01	0.09±0.00
Mg	0.00±0.00	0.00±0.00	0.00±0.00
O	32.00	32.00	32.00

Figure 5. Nepheline composition in CCC and IHT A4 normalized to 32 oxygen as measured through SEM-WDS

Additional characterization of nepheline for impact on glass durability focuses on the calculation of the vacancy concentration. Nepheline is a stuffed derivative of the tridymite polymorph of SiO₂, and the nepheline unit cell is shown in Figure 6 [8]. Tridymite is a tectosilicate (framework silicate) of *cis*-oriented silica tetrahedra with space group P6₃ [8]. Distortion occurs as a result of Al substitution into the $[SiO_4]^{4-}$ tetrahedra found in the tridymite framework [8]. For charge compensation, two of the smaller Na atoms occupy 'oval' rings while the larger K atoms remain in hexagonal rings [8]. In order calculate the vacancies in nepheline from compositional data, a method was implemented to least-squares fit the general formula for a solid solution of components with the compositions (but not necessarily the structures) of anorthite ($\square_{0.5}Ca_{0.5}AlSiO_4$) [An], nepheline ($NaAlSiO_4$) [Ne], kalsilite ($KAlSiO_4$) [Ks], and quartz ($\square Si_2O_4$) [Qz], where $\square$ denotes a vacancy, as described by Blancher et al [8]. The general formula based on the WDS data was hypothesized as:

$$Na_x K_y (Ca+Mg)_{z/2} \square_{w+z/2} (Al+Fe+B)_{x+y+z} Si_{x+y+z+2w} O_{32} \qquad (1)$$

In this modified formula it is assumed that Ca^{2+} and Mg^{2+} enter the Na^+ site resulting in a K^+ site vacancy and that Al, Fe, and B occupy $T(1)$ or $T(4)$ tetrahedral sites [8]. There was not a sufficient concentration of boron in these nephelines to quantify. The merit function minimized for this calculation was D^2 where $D^2=(Na-x)^2+(K-y)^2+((Ca+Mg)-z/2)^2+((Al+Fe+B)-x-y-z)^2+(Si-x-y-z-2w)^2$. The parameters x, y, z, and w represent the fractions of Ne, Ks, An, and Qz [8]. Boundary conditions for the fit were: $x \leq Na$, $y \leq K$, and $z/2 \leq (Ca+Mg)$. The significance of the vacancy concentration is that it serves as an indication of the alkali enrichment in the residual glass. There appears to be a weak temperature dependence of the vacancy concentration (Figure 6); however, since this variance is two orders of magnitude smaller than the concentration of sodium, a major component of the nuclear waste, it may be neglected.

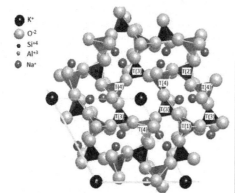

	A4 CCC	A4-750	A4-825
	Average:	Average:	Average:
Na	6.94±0.35	6.89±0.06	6.66±0.06
Al	7.43±0.09	7.39±0.05	7.28±0.03
Si	8.10±0.06	8.14±0.03	8.14±0.03
Fe	0.18±0.08	0.15±0.02	0.16±0.01
Ca	0.31±0.02	0.28±0.02	0.34±0.01
B	0.00	0.00	0.00
K	0.06±0.00	0.09±0.01	0.09±0.00
Mg	0.00±0.00	0.00±0.00	0.00±0.00
O	32.00	32.00	32.00
Vac	0.55	0.58	0.69
D²	0.13	0.18	0.32

Figure 6. (Left) Nepheline structure viewed down [001] with $T(1)$, $T(2)$, $T(3)$, and $T(4)$ tetrahedra types shown, (Right) modelled nepheline composition following equation (1) for A4-CCC, A4-750°C-30 h, and A4-8250°C-30 h where D^2 is the merit function and "Vac" is the vacancy concentration.

CONCLUSIONS

The focus of this work was the investigation of the microstructural and compositional nuances that are associated with the crystallization of high-level nuclear waste simulant glass. It was observed that under a low undercooling condition, nepheline was the most abundant aluminosilicate phase. Upon subjection to increased undercooling, additional phases formed which included calcium phosphate. Continuous cooling via the CCC profile further resulted in the formation of boron-enriched pockets within a tightly-packed nepheline branch network. Elemental analysis revealed that the nepheline composition was independent of thermal history. Postulates were drawn on the significance of this observation in relation to the normative behavior of this melt system.

ACKNOWLEDGMENTS

The WSU NOME laboratory is funded by the Department of Energy's Waste Treatment and Immobilization Plant Federal Project Office contract number DE-EM0002904 under the direction of Dr. Albert A. Kruger. Further support for José Marcial is provided by the Graduate Assistance in Areas of National Need (GAANN) fellowship. Electron microprobe analyses were performed at the Peter Hooper GeoAnalytical Laboratory of the WSU School of the Environment.

REFERENCES

1. J.W. Amoroso, Computer Modelling of High-Level Waste Glass Temperatures within DWPF Canisters during Pouring and Cool Down, SRNL-STI-2011-00546.
2. J.S. McCloy, C.P. Rodriguez, C. Windisch, C. Leslie, M.J. Schweiger, B.J. Riley, J.D. Vienna, Alkali/Alkaline Earth Content Effects on Properties of High-Alumina Nuclear Waste Glasses, Advances in Materials Science for Environmental and Nuclear Technology 63-76 (2010).
3. C.M Jantzen, D.E. Bickford, Leaching of Devitrified Glass Containing Simulated SRP Nuclear Waste. Mat. Res. Soc. Proc. 44 135-146 (1985).
4. T.M. Besmann, K.E. Spear, E.C. Beahm, Assessment of Nepheline Precipitation in Nuclear Waste Glass via Thermochemical Modeling, Mat. Res. Soc. (2000).
5. C.P. Rodriguez, J.S. McCloy, M.J. Schweigher, J.V. Crum, A. Winschell, Optical Basicity and Nepheline Crystallization in High Alumina Glasses, PNNL-20184 (Feb 2011).
6. P. R. Hrma, M.J. Schweiger, B.M. Arrigoni, C.J. Humrickhouse, J. Marcial, A. Moody, C. Rodriguez, R.M. Tate, B. Tincher, Effect of Melter-Feed-Makeup on Vitrification Process, PNNL-18374 (2009).
7. C.M. Jantzen, K.G. Brown, Predicting the Spinel-Nepheline Liquidus for Application to Nuclear Waste Glass Processing. Part II: Quasicrystalline Freezing Point Depression Model, J. Am. Ceram. Soc. 90 1880-1891 (2007).
8. S.B. Blancher, P. D'Arco, M. Fonteilles, M-L. Pascal, Evolution of Nepheline from Mafic to Highly Differentiated Members of the Alkaline series: the Messum Complex, Namibia. Mineralogical Magazine, 74 (3), 415–432 (2010).

Mater. Res. Soc. Symp. Proc. Vol. 1744 © 2015 Materials Research Society
DOI: 10.1557/opl.2015.481

Wet Chemical and UV-Vis Spectrometric Iron Speciation in Quenched Low and Intermediate Level Nuclear Waste Glasses

Jamie L. Weaver[1], Nathalie A. Wall[1], John S. McCloy[2]

[1]Department of Chemistry, Washington State University, Pullman, WA 99164-4630, USA

[2]School of Mechanical & Materials Engineering, Washington State University, Pullman, WA 99164-2920, USA

ABSTRACT

In this study wet chemical methods combined with UV-Vis spectroscopy were performed to quantify $Fe(II)/Fe(III)$ ratios and total iron content of quenched alkali alumino-boro-silicate (simulated nuclear waste) glasses, applying a colorimetric method. We report lessons learned from experimental challenges encountered associated with the colorimetric method, where 1,10 phenanthroline method is complexed with dissolved glass powder and the resulting solution measured for absorbance at 520 nm to determine $Fe(II)$. To obtain total iron, the solution was then equilibrated with a mild reducing agent to chance all Fe to $Fe(II)$, and the absorbance measured again at 520 nm. These absorbance values allowed for calculation of the $Fe(II)/Fe(III)$ ratio, and the total iron content in the glasses. Total Fe measured is somewhat higher than as-batched target values for waste glasses, but very accurate for reference BCR-2G glass. All quenched alumino-boro-silicate glasses analyzed showed a $Fe(II)/Fe(III)$ ratio between 0.06 ($\pm$ 0.01) and 0.04 ($\pm$ 0.01). These values are consistent with those obtained for similar glass compositions melted under analogous conditions, indicating a composition of ca. 94-96% $Fe(III)$.

INTRODUCTION

The disposal of nuclear wastes accumulated in large underground tanks at the Hanford Nuclear Reservation in Richland, WA is a priority goal of the United States Department of Energy (U.S. DOE). DOE plans to first separate the high level waste (HLW) from the low activity waste (LAW), and isolate these wastes in vitrified (glass) waste forms for long term disposal. In the United States, nuclear waste glass (NWG) is primarily based on alumino-boro-silicate glass compositions, chosen for their ability to incorporate a variety of elements, including various multivalent elements potentially present in the nuclear waste slated to be vitrified at the Hanford site in Richland, WA [1]. Of these elements, iron has been found to play an important role in defining the efficiency of the vitrification process, the structure of the glassy matrix, and the long-term durability of the final glasses [2]. Iron present in this glass can be found as a mixture of ferrous or ferric cations [3]. Previous studies have shown that an Fe oxidation state ratio $Fe(II)/Fe(III)$ between 0.1 to 0.5 is ideal for melting conditions [3, 4].

In recent years this $Fe(II)/Fe(III)$ ratio and the total iron content in glass has been studied by Mössbauer, Raman, XANES, and optical spectroscopy methods [5-7]. The first three techniques listed above require access to expensive (both in terms of time and money) and, in the case of XANES, highly specialized equipment which may not be readily available for quick analysis of simulate nuclear waste glasses. However, optical spectroscopy equipment is comparatively inexpensive, relatively quick to execute, and data analysis is straight forward. Many destructive and non-destructive optical spectroscopy procedures have been published for the analysis of iron in glass [8-10]. Obtaining accurate and repeatable values of iron redox and concentration in glasses, in particular, requires robust acid digestion procedures [11, 12] and pH control [13] to avoid oxidation of Fe(II) and provide proper conditions for chromophore complexation. Additionally, care must be taken to avoid reduction of Fe(III) through exposure to light (when complexed with chromophores) [14-16] or grinding materials such as tungsten carbide and alcohol [17]. Despite the numerous caveats to these procedures, the ASTM standard methods either discusses only passingly or without detail glass sample preparation methods, photoreduction, heat evolution with acid additions, and experiment timing, and focus only on the calibration and absorbance calculation [18]. This study therefore presents a detailed experimental protocol that has been assimilated from the previously referenced sources. It also reports lessons learned from experimental challenges encountered during the utilization of wet chemistry methods coupled with UV-Visible spectroscopy to determine the Fe(II)/Fe(III) ratios and total iron content of quenched and non-quenched alkali alumino-boro-silicate, representative of nuclear waste glasses [8].

EXPERIMENTAL DETAILS

A Cary 5000 (Agilent) UV-Vis NIR Spectrophotometry was utilized in this study, and absorption data was obtained with the Cary WinUV software. All samples were measured in transmission mode (%T) at 520 nm in 1 cm plastic cuvettes (3 mL), and referenced to 18 MΩ deionized distilled water (DDIW) at ambient temperature (ca. 20˚C). Standard solutions containing known concentrations of Fe as well as glass sample solutions were run in triplicate.

The simulated NWGs used in this study were provided by collaborators at the Pacific Northwest National Laboratory in Richland, WA [19]. The glasses were batched and melted at PNNL using metal oxides and other typical precursors (e.g., carbonates, boric acid). After batching, each glass was mixed in an agate mill for approximately 3 minutes and then placed into a platinum crucible (Pt/10%Rh) and melted for 1 hour at 1150 - 1400°C, depending on the composition. The resulting glasses were air-quenched by pouring the melt onto a stainless steel plate. Undissolved solids were noted in the glasses after the cooling period ended. To increase the homogeneity, the glass from the first melt was ground to a powder in a tungsten carbide mill (4 minutes, Angstrom Inc.), and melted a second time at the same or slightly lower temperature than the first melt. All the glasses were subsequently air-quenched. The resulting glasses had a uniform appearance upon cooling, and no undissolved solids were observed. The final color of the glasses ranged from dark green to dark brown. Finally, the glasses were ground and sieved to a 100 mesh (149 µm maximum dimension). Before wet digestion of the glass, each powdered sample was placed on a bench top vortex mixer for 2 minutes. It was determined over the course of this study that omission of this step led to inconsistent Fe intensities for replicate samples.

UV-Visible spectrophotometry calibration was established using a set of 10 standard iron solutions prepared from solid ferrous ammonium sulfate hexahydrate (Sigma Aldrich): 0.07 g of the ferrous ammonium sulfate was weighed and transferred into a 1.00 L volumetric flask and diluted with 200.00 mL of DDIW; 3.00 mL of concentrated H_2SO_4 was added, and the resulting solution was diluted with additional DDIW to reach 1 L. 0.0 (method blank); 0.1, 0.5, 1.0, 2.0, 3.0 5.0, 10.0, 25.0, and 50.0 wt % standard iron solutions were prepared following this procedure. These standard solutions were found to be valid for 24 hrs, after which time the color of the solutions would begin to degrade. Thus, the calibration curves and experimental data had to be measured within the same day. To calibrate method accuracy, a BCR-2G reference material (melted and quenched crystal-free " Basalt, Columbia River BCR-2," from the U.S. Geological Survey) was analyzed using the same method as described below with 10 replicates. The experimental wt% of total iron (Fe_{tot} wt%) measured using the procedures detailed below was 9.66 wt% (±0.1, 3σ), equivalent to 13.8±0.1 wt% Fe_2O_3-t or 12.4±0.1 wt% FeO-t, which is well within the reported values for this material [20, 21]. Note, however, that this Fe content is significantly higher than that of the NWGs. The determined Fe(II)/Fe(III) ratio for BCR-2G was 2.02 (±0.2, 3σ) equivalent to Fe(II) 6.45±0.2 wt% or FeO 8.31±0.2 wt%, which compares well with 8.21-8.38 wt% as provided by USGS [22].

0.100 mg of glass was weighed into a 100 mL Teflon beaker, and ca. 0.50 mL of H_2SO_4 (Sigma Aldrich) and ca. 1.00 mL 48 to 51% HF (Sigma Aldrich) were added. The mixture was carefully swirled until all particles dissolved into solution. The cooled solution was then diluted to 100.0 mL with DDIW and well mixed. Using a calibrated pipette, 2.0 mL of the solution was transferred to another 100 mL Teflon beaker, and mixed with an addition of 25 mL of DDIW. An ammonium acetate buffer solution was then used to adjust the pH of the solution to 3.4 ± 0.01, 5.00 mL of 1,10 phenanthroline solution was added to the beaker, and then the solution was diluted to 50 mL with DDIW. 1,10 phenanthroline ($C_{12}H_8N_2$, ortho-phenanthroline or o-Phen, Sigma Aldrich) is a tri-cyclic nitrogen heterocyclic compound that reacts with Fe(II) to form a stable and brilliant orange-colored complex [14]. The estimated time to complete the work described in this paragraph was approximately 5 min per glass sample.

At the beginning of this study it was found that the ratio of HF/H_2SO_4 needed to dissolve the powdered material varied from glass to glass, and that an excess of HF present in solution made it difficult to accurately buffer and dilute the solutions to the desired final volume. Therefore, prior to dissolution of any of the glasses, a set of trial dissolutions were conducted to determine the appropriate ratio of acids needed to dissolve the glass, and the approximate volume of buffer solution required to raise the pH of the solutions to 3.4. The addition of this step improve the overall reproducibility of the experiments.

Ammonium acetate buffer was prepared by dissolving 250.0 mg of ammonium acetate (Sigma Aldrich) to 150 mL with DDIW and adding 700 mL of glacial acetic acid (Sigma Aldrich). The 1, 10 - phenanthroline solution was prepared by dissolving 100.0 mg of 1, 10 phenanthroline monohydrate (Sigma Aldrich) to 100 mL with DDIW, with a few drops of HCl to hasten dissolution. Prior to the addition of the ligand, the solution was placed in a low light atmosphere, and, after ligand addition and final dilution, the solution was placed in a light-tight box until the sample was transferred into the cuvette for analysis to prevent Fe(III)-phen from reducing to Fe(II)-phen [14]. Although not stated in previous protocols, we determined that it was necessary to wait 5 minutes before transferring a portion of the solution to a 1 cm plastic cuvette for analysis. Waiting either more or less than 5 minutes led to variability in the %T measurement. %T was measured three consecutive times for each solution. These measurements gave the Fe(II) concentration.

Within 30 minutes of the first measurement, 30.0 mg of hydroxylamine HCl crystals (a mild reducing agent, Sigma Aldrich) [23] were added to the remainder of the sample, which was then placed on a ca. 90°C hot plate for no more than 4 minutes. The solution was then cooled to room temperature over 12 hours. It was found that allowing the solution to heat longer than 4 minutes or at a temperature higher than 90 °C lead to inconsistent color development in the final product. The resulting solutions were transferred into a clean plastic cuvette and the %T was again measured in triplicate, giving the total Fe concentration. Total ferric iron, Fe(III), concentration was calculated by subtracting the experimentally determined total iron concentration from the ferrous iron, Fe(III), concentration.

RESULTS

The Fe concentration in the glasses were calculated by first completing a linear least squares analysis of the calibration data (%T versus μg Fe_{Total}). The R^2 of the calibration curve calculated in this study for the NWGs was 0.999. The linear equation acquired from this analysis was then used to calculate the micrograms of iron present in the glass based on the recorded %T for the glass.

Iron redox analysis results on the simulated NWG glasses are shown in Table 1. The experimental total Fe wt% (Fe_{Total}) was calculated including three standard deviations (3σ) of error and compared to the reported batch total Fe wt% for the glasses [19]. All glasses analyzed showed a Fe(II)/Fe(III) ratio between 0.06 and 0.04 with an average standard deviation of approximately $\pm$ 0.01 wt%. These values are consistent with that obtained for similar glass compositions melted under similar conditions, indicating ca. 94 – 96% Fe(III) ions [24, 25].

Table 1- Experimentally determination of Fe_{Total}, Wt% Fe_2O_3, and Fe(II)/Fe(III) ratios in HLW glasses. All error values are reported to three standard deviations from the mean of triplicate samples (except reference glass with 10 replicates).

Glass	Wt% Fe_2O_3 (target as-batched [19])	Wt% Fe_2O_3 (total)	Wt% Fe_{Total} measured	Fe(II)/Fe(III)
HLW-E-ANa-13(3Al-3Si)	2.82	3.52 (±0.08)	2.46 (±0.08)	0.06 (±0.02)
HLW-E-ANa-24	3.00	3.52 (±0.09)	2.54 (±0.09)	0.05 (±0.01)
HLW-E-ANa-25	3.00	3.63 (±0.23)	2.83 (±0.23)	0.04 (±0.01)
HLW-E-ANa-26	3.11	3.62 (±0.06)	2.53 (±0.06)	0.04 (±0.02)
IWL-HAC5-1	2.74	3.55 (±0.14)	2.48 (±0.14)	0.06 (±0.01)
BCR-2G (reference)	13.90 [20]	13.80 (±0.1)	9.66 (±0.1)	2.02 (±0.2)

DISCUSSION

It was noted in our analysis of the data that our Fe_{Total} values were consistently higher than the as-batched target. Several sources of potential error are here considered. The actual as-melted overall composition of the glass was never measured by inductively coupled plasma-optical emission spectroscopy (ICP-OES) or similar method, so it is likely that there was some batching error or compositional change from melting. Additionally, some analytical error is possible. The presence of interfering cations, such as Mn^{2+} and Cr^{3+}, in solution that also complex with 1,10 phenanthroline could decrease the total amount of ligand available with which iron could bind [23]. Another source of potential error in the iron measurements can be attributed to inconsistent sampling when the glass was in powder form. A possible error in the Fe(II)/Fe(III) ratios could also occur due to the exposure of the particles to air during milling causing oxidation, or the exposure to tungsten carbide from the mill causing reduction [17].

Future improvements to this procedure are as follows. Using sodium citrate in place of ammonium acetate [18] as a buffer and adding the buffer at the end of the initial sample preparation may help correct the issue of interference ions complexing with the ligand used in this study [23]. Selma and Bandemer have suggested that it is possible to form an iron-phosphate complex in solution, which does not reaction with phenanthroline, if citrate or acetate is added before the ligand [26]. It was also found that the addition of acetate could sometime produce a turbid solution which needed to be centrifuged, while the addition of a citrate buffer produced a clear solution. Although this was not observed in our system, it would be instructive to study the effectiveness of one buffer over the other our system of glasses in a similar manner. This will be the subject of future research with glass of compositions similar to those studied in this method. Accuracy of the measurement could be better validated by measuring a geological standard certified for Fe(II) with lower total Fe concentration that is more similar to that of the NWGs. Finally, an independent confirmation of the total Fe concentration (e.g. with ICP-OES) should be used to eliminate assumption of batching errors and actual composition of the glass.

CONCLUSION

In the assessment of the process and material produced from the vitrification of nuclear waste, it is necessary to quantify the ratio of valence states of elements in the glass matrix. Many methods currently used are time-consuming and require the use of expensive and specialized instrumentation. UV-vis spectroscopy is an exception to these cases; it has been shown to be an effective and quantitative method by which to determine the ratio of Fe(II)/Fe(III) in simulated nuclear waste glasses. With the substitution of a suitable ligand and adjustments to the pH range of experiment, it may be possible to modify the above described procedure for the analysis of other multivalent species in simulated nuclear waste glasses. However, many procedural details have to be considered and carefully controlled. To our knowledge no international standard procedure for assessing Fe in glass exists which provides all the necessary caveats to these measurements, pointing to a need in the community for the establishment of one.

REFRENCES

[1]   B. Grambow, Nuclear Waste Glasses - How Durable?, Elements 2 (6) 357-364 (2006).

[2]   E. Burger, D. Rebiscoul, F. Bruguier, M. Jublot, J. E. Lartigue and S. Gin, Impact of iron on nuclear glass alteration in geological repository conditions: A multiscale approach, Appl. Geochem. 31 (0) 159-170 (2013).

[3]   R. T. Hunter, M. Edge, A. Kalivretenos, K. M. Brewer, N. A. Brock, A. E. Hawkes and J. C. Fanning, Determination of the Fe^{2+}/Fe^{3+} Ratio in Nuclear Waste Glasses, J. Am. Ceram. Soc. 72 (6) 943-947 (1989).

[4]   J. C. Fanning and R. T. Hunter, Nuclear waste glass, and the Fe^{2+}/Fe^{3+} ratio, J. Chem. Educ. 65 (10) 888 (1988).

[5]   I. S. Muller, C. Viragh, H. Gan, K. S. Matlack and I. L. Pegg, Iron Mössbauer redox and relation to technetium retention during vitrification, Hyperf. Inter. 191 (1-3) 347-354 (2009).

[6]   R. Arletti, S. Quartieri and I. Freestone, A XANES study of chromophores in archaeological glass, Appl. Phys. A 111 (1) 99-108 (2013).

[7]   K. Baert, W. Meulebroeck, H. Wouters, P. Cosyns, K. Nys, H. Thienpont and H. Terryn, Using Raman spectroscopy as a tool for the detection of iron in glass, J. Raman Spectrosc. 42 (9) 1789-1795 (2011).

[8]   D. R. Jones, W. C. Jansheski and D. S. Goldman, Spectrophotometric determination of reduced and total iron in glass with 1,10-phenanthroline, Anal. Chem. 53 (6) 923-924 (1981).

[9]   R. Klement, J. Kraxner and M. Liska, Spectroscopic Analysis of Iron Doped Glasse with Composition Close to the E-GLASS: A Preliminary Study, Ceram. Silik. 53 180-183

(2009).

[10]   L. Kido, M. Müller and C. Rüssel, High temperature vis-NIR transmission spectroscopy of iron-doped glasses, Phys. Chem. Glasses 51 (4) 208-212 (2010).

[11]   W. J. French and S. J. Adams, A rapid method for the extraction and determination of iron(II) in silicate rocks and minerals, Analyst 97 (1159) 828-831 (1972).

[12]   L. T. Begheijn, Determination of iron(II) in rock, soil and clay, Analyst 104 (1244) 1055-1061 (1979).

[13]   A. A. Schilt, Analytical Applications of 1,10-Phenanthroline and Related Compounds, Pergamon Press Ltd., Oxford (1969).

[14]   J. W. Stucki and W. L. Anderson, The quantitative assay of minerals for Fe^{2+} and Fe^{3+} using 1,10-phenanthroline. I. Sources of variability, Soil Sci. Soc. Am. J. 45 633-637 (1981).

[15]   J. W. Stucki, The quantitative assay of minerals for Fe^{2+} and Fe^{3+} using 1,10-phenanthroline. II. A photochemical method., Soil Sci. Soc. Am. J. 45 638-641 (1981).

[16]   P. Komadel and J. W. Stucki, Quantitative assay of minerals for Fe^{2+} and Fe^{3+} using 1,10-phenanthroline: III. A rapid photochemical method, Clay Clay Miner. 36 379-381 (1988).

[17]   E. R. Whipple, J. A. Speer and C. W. Russell, Errors in FeO determinations caused by tungsten carbide grinding apparatus, Am. Mineral. 69 987-988 (1984).

[18]   ASTM International, Iron in Trace Quantities Using the 1,10-Phenanthroline Method, E394-09 (2009).

[19]   J. McCloy, N. Washton, P. Gassman, J. Marcial, J. Weaver and R. Kukkadapu, Nepheline crystallization in boron-rich alumino-silicate glasses as investigated by multi-nuclear NMR, Raman, & Mössbauer spectroscopies, J. Noncryst. Solids 409 149-165 (2015).

[20]   K. P. Jochum and U. Nohl, Reference materials in geochemistry and environmental research and the GeoReM database, Chem. Geol. 253 (1–2) 50-53 (2008).

[21]   K. P. Jochum, M. Willbold, I. Raczek, B. Stoll and K. Herwig, Chemical Characterisation of the USGS Reference Glasses GSA-1G, GSC-1G, GSD-1G, GSE-1G, BCR-2G, BHVO-2G and BIR-1G Using EPMA, ID-TIMS, ID-ICP-MS and LA-ICP-MS, Geostand. Geoanaly. Res. 29 (3) 285-302 (2005).

[22]   S. Wilson, USGS: Fe total and Fe(II) values for BCR-2G, by ferrous WDXRF and Fe titration, respectively, Personal communication to: O. Neill, April 29 (2015).

[23]   J. E. Amonette and J. C. Templeton, Improvements to the quantitative assay of nonrefractory minerals for Fe(II) and total Fe using 1,10-phenanthroline, Clay Clay Miner. 46 (1) 51-62 (1998).

[24]   B. Cochain, D. R. Neuville, G. S. Henderson, C. A. McCammon, O. Pinet and P. Richet, Effects of the Iron Content and Redox State on the Structure of Sodium Borosilicate Glasses: A Raman, Mössbauer and Boron K-Edge XANES Spectroscopy Study, J. Am.

Ceram. Soc. 95 (3) 962-971 (2012).

[25]   R. K. Kukkadapu, H. Li, G. L. Smith, J. D. Crum, J.-S. Jeoung, W. Howard Poisl and M. C. Weinberg, Mössbauer and optical spectroscopic study of temperature and redox effects on iron local environments in a Fe-doped (0.5 mol% Fe_2O_3) $18Na_2O-72SiO_2$ glass, J. Non-Cryst. Solids 317 (3) 301-318 (2003).

[26]   S. L. Bandemer and P. J. Schaible, Determination of Iron. A Study of the o-Phenanthroline Method, Ind. Eng. Chem. Analy. Ed. 16 (5) 317-319 (1944).

Mater. Res. Soc. Symp. Proc. Vol. 1744 © 2015 Materials Research Society
DOI: 10.1557/opl.2015.496

A Sampling Method for Semi-Quantitative and Quantitative Electron Microprobe Analysis of Glass Surfaces

Jamie L. Weaver, Joelle Reiser, Owen K. Neill, John S. McCloy, Nathalie A. Wall.

Washington State University, Chemistry Department, Pullman, WA 99164, USA

ABSTRACT

The determination of the long-term stability and corrosion of vitrified nuclear waste is an important aspect of research for the U.S. Department of Energy (DOE). It is necessary to understand the rate and mechanisms of Nuclear Waste Glass (NWG) corrosion to determine whether or not the glassy matrix will be able to retain radionuclides for the required repository performance time period. Glass corrosion and the rate of glass corrosions is determined by both chemical and microscopy. Electron Microprobe Analysis (EPMA) is a common and powerful method utilized in the examination of the chemographic difference between corroded and uncorroded NGWs. In this work, two forms of quantitative and semi-quantitative EPMA methods are defined by optimizing the instruments counting statistics against a standard glass and NIST minerals that have compositions similar to the glasses under examination. Data collected on both the planar and cross-sectioned surfaces of an unaltered simulated NWG by Standard based Wavelength Dispersive Spectroscopy (WDS) was found to be comparable to the theoretical composition of the glass. Conventional standardless Energy Dispersive Spectroscopy (EDS) data collected on the same surfaces was not comparable. However, standard-based EDS analysis is shown to be able to discriminate between unaltered and corroded glass surfaces.

INTRODUCTION

Energy Dispersive Spectroscopy-Electron Probe Micro Analysis (EDS-EPMA) is a common method used to analyze corroded glass surfaces. In the past EDS-EPMA has been used to identify the presence of a hydrated and silica-rich alteration layer and the formation of solid composites on the glass surface [1-3]. EDS allows the user to obtain clear and detailed images of the corroded surface morphology over large area (μm^2), and to characterize the element(s) involved in the glass corrosion process. Glass studies often couple EPMA with Induction Coupled Plasma – Optical Emission Spectrometry (ICP-OES) to determine glass corrosion rate, and to attempt definition of glass corrosion mechanisms [4-6]. Most of these studies utilize "standardless" EDS, a method that suffers from both accuracy and precision limitations, especially in comparison to wavelength-dispersive X-ray spectroscopy (WDS) [7]. This form of EDS is sufficient to detect the gross chemical variations caused by the alteration processes, such as complete loss of Na in the altered layers, but cannot accurately quantify the amount of loss. ICP-OES data is used to fill in this information. However, in situations when it is not possible to obtain or analyze a sample of the corroding solution, analysis of a corroded surface by EPMA becomes more important, and more rigorous methods are required.

One such method is "standard" based EDS, which is semi-quantitative. In this method an unaltered sample of the glass is measured with the corroded samples under the same analysis conditions, and then the calculated elemental weight percent's (wt%) of the altered glasses are compared to those of the unaltered glass. A second method that can be used is standard based WDS. The setup for this form of analysis is similar to that used for the standard based EDS method, but it includes an additional calibration setup that compares the X-ray intensities collected from the unaltered and corroded glasses to a set of standard samples. In this study both methods were executed on sets of unaltered and altered glass samples. The sets of samples were measured in a planar orientation, and in cross-section.

EXPERIMENTAL DETAILS

International Simple Glass (ISG was chosen for this study due to its relatively simple oxide composition (Table 1). A 50 kg ISG sample was produced in May 2012 by MoSCI Corporation (Rolla, MO) and subsequently divided into 500 g glass ingots.[4] The glass ingot used in this study was cut into coupons with an oil (IsoCut™ Fluid, Buehler) lubricated diamond saw (IsoMet™ Low Speed Saw, Buehler). Approximate dimensions of the coupons were 30 mm × 10 mm × 4 mm. Each coupon was polished at 120, 320, 600, and 1200 grits using an oil-based lubricant (IsoCut™ Fluid, Buehler) on a Leco Grinding and Polishing Polisher.

Element	Wt%
Si	56.20
B	5.37
Na	9.05
Al	3.23
Ca	3.57
Zr	2.44

Table 1. ISG elemental composition by weight percent [4].

Table 2 lists the type of EPMA analysis used on each sample. The $MgCl_2$ and water corroded samples were prepared in the following manner: ISG coupons were prepared in the above described way and then subjected to 7 day Product Consistency Tests (PCT) as defined in ASTM C1285 [8]. These tests were conducted at 90˚C in distilled deionized water (18MΩ) or saturated $MgCl_2$ solution. The surface area/volume value for these samples was 2.5 (± 0.1) cm^3/mL solution. At the end of the test the samples were carefully removed from the corroding solution and dried under ambient conditions for 48 hours before being placed in a desiccator for storage.

Coupons designated to be measured in planar orientation were lightly coated with a layer of carbon and grounded to the sample holder prior to being placed in the EPMA. The unaltered coupons to be measured in the planar orientation were oil polished with 6 μm diamond paste prior to being carbon coated. Coupons chosen to be measured in cross-section were first diagonally cut with a tungsten carbide blade, placed on end in ~2 mL of Specifix-20 resin, and allowed to try for 20 to 24 hours. The cross-sectioned surface was then polished to a smooth using 9 μm diamond paste (Ted Pella), and then carbon coated and grounded to the sample holder with carbon tape.

All EPMA analyses were conducted using the JEOL JXA-8500F electron microprobe at the Peter Hooper GeoAnalytical Lab of the WSU School of the Environment, equipped with 5 WDS spectrometers, a Thermo Scientific UltraDry EDS spectrometer, ProbeForEPMA WDS automation and analytical software for WDS analyses, and ThermoNSS System7 automation and analytical software for EDS analyses. EDS measurements were made using an accelerating voltage of 15 kV, and a beam current of 8 nA. The beam was defocused to a 1 μm spot size to help mitigate alkali migration under beam irradiation. EDS spectra were collected for a total of 60 live-time seconds. Elemental concentrations from these measurements were determined from raw EDS intensities using the PROZA $\phi(\rho z)$ matrix correction, similar to that outlined in Bastin and Heijligers, through the ThermoNSS System 7 software package [9].

Sample	Unaltered, planar orientation	Corroded with MgCl$_2$, Planar orientation	Corroded with H$_2$O, Planar orientation	Unaltered, cross-section orientation
Analyses Type	EDS/WDS	EDS	EDS	EDS/WDS

Table 2. Type of analyses (EDS or WDS) and samples analyzed.

WDS measurements were made using an accelerating voltage of 15 kV, a beam current of 6 nA, and a beam size of 10 μm. A list of peak and background counting times and calibration standards can be found in Table 3. Concentrations were determined from raw X-rays measured by WDS using a ZAF intensity correction (Armstrong-Love/Scott) through ProbeForEPMA analytical software.

Analysis of the planar surface of an unaltered ISG coupon was conducted by dividing the surface into a 3 x 5 grid (Figure 1). One data point was taken for each grid square for a total of fifteen points being collected per planar surface. Alternatively, the cross-sectioned surfaces were divided into 1 x 5 grid, and three data points were taken per each square. The total number of data points collected from the cross-sectioned surfaces was also fifteen. Although EPMA information was collected on all oxides in ISG, only three are discussed here. These three oxides were chosen because they are known to be leached

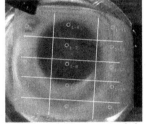

Figure 1. Grid overlaid on uncorroded planar oriented ISG glass coupon. The red marks show the approximate areas where EDS and WDS measurements were acquired on the surface.

during the corrosion of ISG [10]. Boron, an element commonly used to monitor the corrosion process, has been emitted from this list due to complications with measuring it with EPMA [11].

Statistical analysis of the data was conducted using the Statistical Package for the Social Sciences (SPSS) Software 21 (IBM). Contrary to its name, SPSS is a powerful software package widely used in all scientific fields. Wt% data collected from EPMA analysis of all surfaces was imported into SPSS, and then group statistical tests (calculation of mean, standard deviation, and analysis and removal of outliers), Independent Samples T-tests, and Levene's Test for Equality of Variances. Statistical group analysis was completed for all measured surfaces. T-tests and Levene's Tests were conducted on the EDS and WDS measurements taken from the unaltered planar ISG sample and the unaltered cross-sectioned ISG sample. Levene's test values were calculated for each set of comparison, and were found to be significant for all t-tests.

Element	Peak Counting Time (s)	Low Background Counting Time (s)	High Background Counting Time (s)	Spectrometer #	Analyzing Crystal	Calibration Standard
SiO_2	20	10	10	4	PETJ	Kakanui Anorthoclase (USNM 133868)
Al_2O_3	15	7.5	7.5	1	TAP	ISG
Na_2O	15	7.5	7.5	1	TAP	Wilburforce Hornblende

Table 3. Details of WDS measurement conditions, including beam conditions, peak and background counting times and calibration standards

DISCUSSION

Group statistical data for all EDS analyses are highlighted in Table 4. Measurement errors were calculated to two standard deviations. Calculated elemental wt% for corroded surfaces showed large errors as compared to the unaltered surfaces. This is most likely due to uneven corrosion across the surface of the coupons. The unaltered surfaces did not show as large of error values as the corroded surfaces, but t-test values show that their means differ significantly (Table 4). This was as unexpected result as it was assumed that the composition of the unaltered planar surface was the same as the unaltered cross-sectioned sample. The elemental wt% values were also not in good agreement with the theoretical wt% values for ISG (table 1). To investigate whether this difference was the result of method error or from compositional differences between the two samples the two surfaces were measured with WDS.

WDS results show significantly less difference between the mean values measured for the two surfaces, and both values were in good agreement with theoretical wt% for each oxide. The calculated errors for the WDS measurements were also smaller except in the case of SiO_2. Therefore, our two unaltered glass surfaces are statistically similar.

Oxide	Unaltered, planar orientation (wt%)	Corroded with $MgCl_2$, planar orientation (wt%)	Corroded with H_2O, planar orientation (wt%)	Unaltered, cross-section orientation (wt%)
Na_2O	10.6, 1.8	2.0, 2.7	1.79, 3.0	5.0, 2.4
A_2O_3	3.9, 0.2	3.8, 1.9	6.10, 2.2	4.2, 0.6
SiO_2	32.0, 1.0	21.8, 14.5	33.31, 2.3	35.0, 1.0

Table 4. Group statistics for samples measured by EDS. The first number listed is the mean of all point measured, and the *second number* is the error of the measurement (2σ, 95% confidence interval).

CONCLUSION

Standard based EDS analysis of the unaltered and corroded surfaces of a simulated nuclear waste glass has been shown to be able to detect gross chemical variations caused by the alteration processes. However, at this time, it cannot accurately quantify the amount of elements leached from a corroded glass surface. Elemental wt% calculated from standard based WDS analysis of two surfaces of unaltered glasses has been shown to be in good agreement with theoretical elemental wt% values. Additional research is being conducted to determine whether this method of WDS can be used to accurately determine the difference in elemental composition between a corroded an unaltered glass surface.

EDS		Planar Surface	Cross-Section	WDS		Planar Surface	Cross-Section
Na (wt%)	mean	10.6	5.0	Na (wt%)	mean	9.1	9.1
	σ	1.8	2.4		σ	0.1	0.2
	t-test	18.7			t-test	0.2	
	p	7.6			p	1.3	
	df	27			df	24	
Al (wt%)	mean	3.9	4.2	Al (wt%)	mean	3.3	3.3
	σ	0.2	0.6		σ	0.03	0.03
	t-test	-3.6			t-test	-0.6	
	p	6.3			p	0.9	
	df	27			df	24	
Si (wt%)	mean	32.0	35.0	Si (wt%)	mean	26.2	26.2
	σ	1.0	1.0		σ	0.1	0.1
	t-test	-21.8			t-test	1.0	
	p	7.6			p	0.1	
	df	27			df	24	

Table 5. Results of statistical analysis of WDS measurements of planar and cross-sectioned unaltered ISG surfaces. All error is reported to two standard deviations measurement (2σ, 95% confidence interval).

ACKNOWLEDGMENTS
This research is being performed using funding received from the DOE Office of Nuclear Energy University Programs, under Project 23-3361. The authors would like to thank Dr. Owen Neill of the Peter Hooper GeoAnalytical Lab of the WSU School of the Environment for his assistance with the electron microprobe solid state analysis.

REFERENCES

[1]   M. Melcher, R. Wiesinger, and M. Schreiner, Acc. Chem. Res. 43, 916-926 (2010).

[2]   P. Bellendorf, G. Gerlach, P. Mottner, E. López, in *Glass & Ceramic Conservation 2010*, edited by H. Roemich, (Inter. Coun. Muse. Symp. Proc. Corning, New York, Jun 2010), 137-143.

[3]   P. Bingham and C.M. Jackson, J. Archaeol. Sci. 35, 302–309 (2008).

[4]   S. Gin, A. Abdelouas, L.J. Criscenti, W.L. Ebert, K. Ferrand, T. Geisler, M.T. Harrison, Y. Inagaki, S. Mitsui, K.T. Mueller, J.C. Marra, C.G. Pantano, E.M. Pierce, J.V. Ryan, J.M. Schofield, C.I. Steefel, J.D. Vienna, Mater. Today. 16, 243–248 (2013).

[5]   J.P. Boudreault, J.S. Dubé, M. Sona, E, Hardy, E. Sci. Total Environ. 425, 199–207, (2012).

[6]   T. Geisler, A. Janssen, D. Scheiter, T. Stephan, J. Berndt, A. Putnis, J. Non. Cryst. Solids. 356, 1458–1465 (2010).

[7]   D. E. Newbury, *Microsc. Microanal.* 11, 545–561 (2005).

[8]   ASTM C1285. Annual Book of ASTM Standards, Vol. 12.01: Standard Test Methods for Determining Chemical Durability of Nuclear, Hazardous, and Mixed Waste Glasses and Multiphase Glass Ceramics: The Product Consistency Test (PCT)

[9]   G. F. Bastin, J. M. Dijkstra, and H. J. M. Heijligers, X-Ray Spectrom. 27, 3-10, (1998).

[10]   Y. Inagaki, T. Kikunaga, K. Idemitsu, and T. Arima, Int. J. Appl. Glas. Sci. 4, 317–327 (2013).

[11]   M. Chaussidon, F. Robert, D. Mangin, P. Hanon, and E.F. Rose, Geostandards Newsletter 21, 7–17 (1997).

Mater. Res. Soc. Symp. Proc. Vol. 1744 © 2015 Materials Research Society
DOI: 10.1557/opl.2015.311

The void fraction of melter feed during nuclear waste glass vitrification

Zachary J. Hilliard[1] and Pavel R. Hrma[1, 2]

[1]Pacific Northwest National Laboratory, 902 Battelle Blvd,
Richland WA, 99354, U.S.A.

[2]Division of Advanced Nuclear Engineering, Pohang University of Science and Technology,
Pohang 790-784, Korea

ABSTRACT

To efficiently vitrify Hanford waste, the melting process (i.e., melter feed turning into waste glass) must be modeled and optimized. The rate of heat transfer to the melter feed in a waste glass melter, and thus the rate of melting, is strongly affected by the melter feed porosity, especially in the final stages where the glass-forming melt produces foam that insulates the feed from the molten glass. The volume expansion test allows the determination of the melter feed porosity as a function of temperature. This test measures the profile area of the feed pellet as it turns into glass. This contribution presents the calculation of the void fraction (porosity) of the melter feed as a function of temperature, heating rate, and material parameters. The process of finding the void fraction is described as well as results from the application of this process.

INTRODUCTION

To efficiently vitrify Hanford waste [1], the melting process (i.e., melter feed turning into waste glass) must be modeled and optimized [2,3]. The rate of heat transfer [4] to the melter feed in a waste glass melter, and thus the rate of melting, is affected by the melter feed porosity, especially in the final stages where the glass-forming melt produces foam that insulates the feed from the molten glass [5,6]. The volume expansion test (also dubbed the "pellet test" because it uses small cylindrical pellet samples) allows the determination of the melter feed porosity as a function of temperature, heating rate, and material parameters [7,8].

For this test, a feed pellet of known composition, mass, and dimensions is placed on an alumina disk inside a furnace and is fitted with thermocouples. Pictures are taken of the pellet profile at various temperatures as the pellet is heated from room temperature to approximately 1000°C. From these pictures, each of which corresponds to a unique temperature, the void fraction of the pellet must be determined.

The void fraction, ϕ, is the ratio of the void volume, V_v, to the bulk volume, V_b, i.e., $\phi = V_v/V_b$. This expression can be written in terms of the pellet mass:

$$\phi(T) = 1 - \frac{m(T)}{\rho(T)V_b(T)} \tag{1}$$

where ρ is the material density of the sample and T is the heat treatment temperature.

Because the rate at which the sample is heated can affect various parameters such as weight loss as a function of temperature, kinetic models that account for the change in heating rate should be used when possible. Kinetic models for the mass loss fraction [9,10], which is the ratio of the mass at some temperature, T, to the original sample mass, and quartz fraction [11,12], which will be discussed later, have been developed and will be implemented here. Because the original sample mass is known, this leaves $\rho(T)$ and $V_b(T)$ to be determined. This paper presents a model to estimate the void fraction of a heated pellet by developing an algorithm to calculate $V_b(T)$ from profile images and a mathematical model to estimate $\rho(T)$ based on the measured density of rapidly quenched heat-treated feed samples.

BULK VOLUME

The pellet volume must be determined from its profile area using photographic images taken during the pellet heat treatment. Previously this was done by sectioning the pellet profile into thin segments and, assuming that the pellet is a body of revolution, summing the volumes of the corresponding cylindrical discs. A rod gage was used for scale [8]. This procedure has been improved using a computer program capable of increasing the number of elements and computing the volume in an efficient manner.

We use MATLAB's photo processing capability to determine the pellet boundary. A one pixel vertical spacing allows for the best possible numerical volume approximation (Figure 1). The boundary is split into left and right "sub-boundaries", each containing the same number of points that span the same vertical displacement.

Figure 1. Boundary pixels defined by MATLAB. In this instance, the picture was manually edited to make a more distinctive boundary for MATLAB to detect.

Points are defined at each boundary pixel so that trapezoidal profile elements are formed. Volume elements are then constructed by letting each trapezoid be the profile of a truncated slanted cone and taking that truncated slanted cone to be the volume element. The trapezoid itself is taken to be an area element.

DENSITY CHANGE DURING QUENCHING

The quenched material density, $\hat{\rho}$, can be measured at various heat treatment temperatures using gas pycnometry so that a function $\hat{\rho}(T)$ can be found by fitting a curve to the data. Because the quenched density is greater than the density when the sample was at the heat treatment temperature due to volume contraction, a method to estimate the material density of the sample at the heat treatment temperature must be constructed. If we assume that at temperatures equal to or greater than some temperature T_m where the glass-forming phase fully connects (marked by a local minimum in the volume-temperature plot) the sample consists of only two phases, then ρ can be calculated from $\hat{\rho}$ by calculating the change in density due to volume contraction of each phase individually. The first phase is quartz which melts into the second phase (called melt) that is composed of all non-solid components. Thus, the only significant composition change that occurs in the feed at or above T_m as it is heated is the dissolution of quartz in the glass-forming phase. There is no mass lost from the sample due to the evolution of gas above T_m (the mass of gases from collapsing foam is negligible). The quartz does not undergo any physical changes that affect either its thermal expansion properties or its quenched density. The glass-forming phase undergoes the glass transition at the glass-transition temperature, T_g, at which the volume expansion abruptly changes, but the density remains a continuous function of temperature (there is no step change that occurs at the melting or freezing point of a crystalline substance). The T_g and expansion coefficients of molten glass (the glass-forming phase) below and above the T_g change as the melt composition changes in response to the dissolving quartz. Hence, the only composition change is the increase of silica fraction from nearly zero to the final fraction of the homogeneous glass with quartz completely dissolved. Here we use the average composition of the melts, ignoring nonuniformity associated with the diffusion-driven dissolution and homogenization.

Using for a glass property-composition relationship the linear function, $p = \sum_N p_i g_i$ where g_i is the ith component mass fraction and p_i is the ith component partial specific property, we can obtain the relationship

$$p(T) = \frac{p^G - p^s c_Q(T)}{1 - c_Q(T)} \tag{2}$$

The thermal volumetric expansion coefficient is defined as $\alpha \equiv \frac{1}{V}\frac{dV}{d\theta}$. For $T < T_g$, $\alpha = \alpha_S$, where α_S is the solid glass thermal expansion and for $T > T_g$, $\alpha = \alpha_L$, where α_L is the molten glass thermal expansion. Using these values, the melt density can be approximated as

$$\rho_M(T) = \hat{\rho}_M(T) \exp\left[\begin{matrix} \alpha_S(T)(T_0 - T) + \\ \left(\alpha_S(T) - \alpha_L(T)\right)\left(T - T_g(T)\right)\mathcal{H}\left(T - T_g(T)\right) \end{matrix} \right] \tag{3}$$

where the functions $\alpha_S(T)$, $\alpha_L(T)$, and $T_g(T)$ have the form of Eq. (8); $\hat{\rho}_M(T)$ can be found using the known function $\rho_0 = \hat{\rho}(T)$; and $\mathcal{H}$ is the Heaviside step function. In this simplified model, we ignore the change in the crystal structure of quartz and assume that only volume expansion affects quartz density between T_0 and T.

The material properties are shown in Table 1 and the quenched density is shown in Figure 2 [13]. The calculation was performed for the temperature interval from T_m = 820°C (determined by the temperature at which the minimum volume occurred, see Figure 3) to 1200°C. For the temperature interval from T_0 = 100°C to T_m, the density was estimated using linear interpolation. The resulting $\rho = \rho(T)$ function is plotted in Figure 2 alongside $\rho_0 = \hat{\rho}(T)$ for comparison.

RESULTS

The void fractions, shown in Figure 3, were calculated using the data given in Table 1, the method for calculating the bulk volume given in section above titled "density change during quenching", and the kinetic mass loss model developed from TGA data gathered from A0 [8,13] feed. The normalized area and volume plots are consistent with previous work [8], and the void fraction plot makes sense when considering both the bulk volume trend and the density trend.

Table 1. Material property values used in the calculation of density

Data			Reference
g_S^G	--	0.305	[9]
α_S^G	K^{-1}	3.76E-5	[14]
α_L^G	K^{-1}	2.40E-4	[14]
α_S^s	K^{-1}	4.90E-6	[14]
α_L^s	K^{-1}	1.71E-4	[14]
T_g^G	°C	469	[15]
T_g^s	°C	634	[15]
α_Q	K^{-1}	1.75E-5	[16]
$\hat{\rho}_Q$	$g\ cm^{-1}$	2.648	[17]

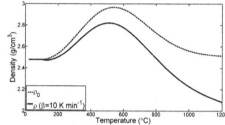

Figure 2. Density for A0 heated at 10 K min^{-1}

From 25°C to 600°C, the bulk volume slightly decreases while the density increases and the mass decreases by approximately 20% (Figure 3). The density increase in combination with the mass decrease results in a significant decrease in material volume so that the void fraction increases to a maximum in the neighborhood of of 600°C. From 600°C to 800°C, the void fraction decreases because the bulk volume decreases as the material volume increases due to a decrease in density and very little mass loss. From 800°C to approximately 900°C, the bulk volume rapidly increases due to the entrapment of gases in the connecting feed. The effect of this on the void fraction is slightly offset by a decrease in density, but the increase in bulk volume due to the trapping of evolved gas is much greater than the increase in material volume due to volume expansion so that the void fraction increases in much the same manner as the bulk volume. At 900°C, the rate that the evolved gas escapes the glass (due to the pressure of the trapped gas being great enough to overcome both the atmospheric pressure and glass melt viscosity) exceeds the rate at which gas is being trapped so that the bulk volume begins to decrease which causes the void fraction to follow suit.

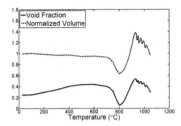 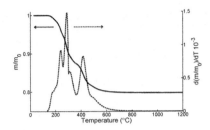

Figure 3. Left: Final results for A0 feed at 10 K min⁻¹ Right: Mass loss for 10 K min⁻¹

CONCLUSIONS

Determining the evolution of the void fraction during the conversion of Hanford waste melter feeds to glass allows further progress to be made to optimize the melting process. As more data is collected, the void fraction model can be further generalized to better account for the heating rate and melter feed composition as in addition to temperature.

The MATLAB program that uses the density model as described in this paper allows the void fraction of feed samples to be calculated at various temperatures so that the void fraction can be estimated as a function of temperature. Once additional data is taken so that the effect of the heating rate (on the bulk volume, mass loss, and density) can be determined with greater accuracy, the void fraction can be estimated as a function of temperature and heating rate.

ACKNOWLEDGMENTS

David Pierce, Brad Vanderveer: data collection

REFERENCES

1.   R. A. Kirkbride, G. K. Allen, R. M. Orme, R. S. Wittman, J. H. Baldwin, T. W. Crawford, J. Jo, L. J. Fergestrom, T. M. Hohl, D. L. Penwell, Tank waste remediation system operation and utilization plan, Vol. I, HNF-SD-WM-SP-012, Numatec Hanford Corporation, Richland Washington (1999).
2.   R. Pokorny, P. Hrma, Mathematical modeling of cold cap, J. Nucl. Materials 429, 245-256 (2012).
3.   R. Pokorny, P. Hrma, Model for the conversion of nuclear waste melter feed to glass, J. Nucl. Materials 445, 190-199 (2014).
4.   J. A. Rice, R. Pokorny, M. J. Schweiger, P. Hrma, Determination of heat conductivity and thermal diffusivity of waste glass melter feed: Extension to high temperatures, J. Am. Ceram. Soc., 1-7 (2014).
5.   P. Hrma, Melting of Foaming Batches: Nuclear Waste Glass, Glastech. Ber. 63K, 360-369 (1990).

6. R. Pokorny, A. A. Kruger, P. Hrma, Mathematical modeling of cold cap: Effect of bubbling on melting rate, Ceram. Silikaty 58, 296-304 (2014).
7. P. Hrma, M. J. Schweiger, C. J. Humrickhouse, J. A. Moody, R. M. Tate, T. T. Rainsdon, N. E. TeGrotenhuis, B. M. Arrigoni, J. Marcial, C. P. Rodriguez, B. H. Tincher, "Effect of glass-batch makeup on the melting process," Ceramics-Silikaty 54, 193-211 (2010).
8. S. H. Henager, P. Hrma, K. J. Swearingen, M. J. Schweiger, J. Marcial, N. E. TeGrotenhuis, Conversion of batch to molten glass, I: Volume expansion, J. Non-Cryst. Solids 357, 829-835 (2011).
9. R. Pokorny, D. A. Pierce, P. Hrma, Melting of glass batch: Model for multiple overlapping gas-evolving reactions, Thermochimica Acta 541, 8-14 (2012).
10. C. Rodriguez, J. Chun, M. Schweiger, P. Hrma, Understanding cold-cap reactions of nuclear waste feeds through evolved gas analysis, Thermochimica Acta 592, 86-92 (2014).
11. P. Hrma, K. J. Swearingen, S. H. Henager, M. J. Schweiger, J. Marcial, N. E. TeGrotenhuis, Conversion of batch to molten glass, II: Dissolution of quartz particles, J. Non-Cryst. Solids 357, 820-828 (2011).
12. P. Hrma, J. Marcial, Dissolution retardation of solid silica during glass-batch melting, J. Non-Cryst. Solids 357, 2954-2959 (2011).
13. J. Marcial, J. Chun, P. Hrma, M. J. Schweiger, Effect of bubbles and silica dissolution on melter feed rheology during conversion to glass, Environ. Sci. and Technol. 48, 12173-12180 (2014).
14. P. Hrma, G. F. Piepel, M. J. Schweiger, D. E. Smith, D. S. Kim, P. E. Redgate, J. D. Vienna, C. A. LoPresti, D. B. Simpson, D. K. Peeler, M. H. Langowski, Property/Composition Relationships for Hanford High-Level Waste Glasses Melting at 1150°C, PNL-10359, Vol. 1 and 2, Pacific Northwest Laboratory, Richland, Washington (1994).
15. P. Hrma, Glass viscosity as a function of temperature and composition: A model based on Adam-Gibbs equation, J. Non-Cryst. Solids 354, 3389-3399 (2008).
16. Y. Linard, H. Nonnet, T. Advocat, Physicochemical model for predicting molten glass density, J. Non-Cryst. Solids 354, 4917-4926 (2008).
17. "Physical Constants of Inorganic Compounds", in CRC Handbook of Chemistry and Physics, Internet Version 2005, David R. Lide, ed., <http://www.hbcpnetbase.com>, CRC Press, Boca Raton, FL, 2005.

Mater. Res. Soc. Symp. Proc. Vol. 1744 © 2015 Materials Research Society
DOI: 10.1557/opl.2015.395

Charge Compensation in Trivalent Doped Ca$_3$(SiO$_4$)Cl$_2$

M. R. Gilbert
AWE, Aldermaston, Reading, RG7 4PR, UK.

ABSTRACT

Calcium chlorosilicate (Ca$_3$(SiO$_4$)Cl$_2$) is seen as a potential host phase for the immobilization of Cl-rich wastes arising from pyrochemical reprocessing, a waste stream often containing a mix of both di- and trivalent cations. Substitution of trivalent cations into the lattice requires some form of charge compensation to ensure the lattice remains charge neutral overall. Whilst previous work has only examined this through the formation of Ca vacancies, this study investigates the feasibility of charge-balancing via the substitution of a monovalent cation onto the Ca sites of the lattice. To that end, a series of static lattice calculations were performed to determine the site selectivity of monovalent cations of differing size when substituted onto the Ca sites of the calcium chlorosilicate lattice and the solution energies for the overall substitution processes compared with those for charge compensation via vacancy formation. In all cases the monovalent charge-balancing species shows a clear preference for substitution onto the Ca1 site in the calcium chlorosilicate lattice. The solution energy of the substitution process increases with the increasing ionic radii of both the mono- and trivalent species as the steric stresses associated with substitution of larger cations than the Ca^{2+} host increase. As such, only charge-balancing using Li$^+$, Na$^+$ or K$^+$ is more favourable than via formation of a Ca vacancy.

INTRODUCTION

Calcium chlorosilicate (Ca$_3$(SiO$_4$)Cl$_2$) has one of the highest Cl contents of any chlorine-containing silicate at ~ 25 wt. % [1]. It is monoclinic in structure, with space group P2$_1$/c, containing 4 formula units per unit cell [2]. The structure consists of an approximately cubic close-packed arrangement of SiO$_4$$^{4-}$ tetrahedral and Cl$^-$ anions, with the Ca^{2+} cations occupying interspersed octahedral sites in such a way that a distorted NaCl-type structure results [3].

In calcium chlorosilicate the wastes can be directly incorporated into the lattice structure via alter- or isovalent substitution. The potential for calcium chlorosilicate to incorporate cations in this manner has already been demonstrated for transition metal and rare earth elements in the production of photoluminescent materials [4]. Direct incorporation within a ceramic lattice in this way allows a much more precise crystallo-chemical understanding of the incorporation method and will provide a far superior level of chemical immobilization compared to physical encapsulation within a glass or cement waste-form. In addition, using a synthetic mineral composition analogous to that found in nature can provide an indication of the potential long-term behaviour of the waste-form.

In a previous study, the site selectivity of both single and multiple di- and trivalent dopant cations substituted onto the three calcium sites within the calcium chlorosilicate unit cell was investigated [5]. In that work, the only charge-compensation mechanism considered to keep the lattice charge neutral when doped with trivalent cations was the creation of a calcium vacancy. Here, the possibility of charge-compensation via substitution of a charge-balancing species (in this case a monovalent cation) is examined.

METHODOLOGY

Throughout this work the GULP (General Utility Lattice Program) code has been used [6], which enables the simulation of both perfect and defective lattice structures by modelling the potential energy of the system as a function of its atomic coordinates [7]. This enables the calculation of the total lattice energies of structures, subject to energy minimisation via the BFGS algorithm [8]. This approach treats the materials as fully ionic, using the Ewald technique to sum the electrostatic interactions between each of the formally charged species [9]. The two-body, short range interactions within the crystal structure, are modelled using Buckingham potentials. Ionic polarisability is incorporated into the model using the Dick and Overhauser shell model [10], which treats each ion as a massless charged shell of valence electrons interacting with an inner core via a harmonic potential. It is also important that any deviations from the SiO_4 tetrahedral arrangement are correctly described and that any changes to the O–Si–O bond angle incur the appropriate energy penalty, therefore an O–Si–O three-body bending term is also included.

Defect calculations are based upon the Mott-Littleton approximation [11]. In this approach the crystal is divided into concentric spherical regions about a central point (usually taken to be the defect or mid-point between multiple defects), termed regions I, IIa and IIb, in which progressively more approximate methods are used to calculate the lattice response to the defect or defects. In region I, the central inner region surrounding the defect, all interactions are treated at an atomistic level and the ions are explicitly relaxed under the perturbation generated by the defect. In region IIa, the greater distance from the defect allows the ions to be treated by more approximatly, as it is assumed that the only relevant feature of the defect for ions beyond the inner region is its charge with respect to the lattice. The outer region, IIb, extends to infinity by virtue of the Ewald summation technique and so can effectively be considered as an array of point charges at the perfect lattice sites [12].

The potentials used in this work have been taken from the work of Dove on a transferable model for silicate minerals [13], with the potentials for chlorine and defect species those derived by Rabone and de Leeuw [14].

RESULTS AND DISCUSSION

Monovalent Substitution

Initially, isolated substitutions of monovalent cations onto each of the three Ca sites were considered, in order to determine any degree of site selectivity on the part of the monovalent cations. Such a substitution process requires the creation of a Cl vacancy in order for the lattice to remain charge neutral. As such, it can, for the purposes of comparing substitutional site selectivity, be considered to proceed via the equation:

$$ACl + Ca^x_{Ca(x)} + Cl^x_{Cl(x)} \rightarrow A^/_{Ca(x)} + V^/_{Cl(x)} + CaCl_2 \tag{1}$$

where A is the monovalent dopant cation and $V^/$ the vacancy formed.

The defect energies resulting from the formation of a vacancy on each of the Cl sites are 3.29 eV and 3.27 eV for the Cl1 and Cl2 sites respectively. The defect energies for each site are almost identical, with a slight favouring of the Cl2 site.

Monovalent cations of a range of ionic radii were substituted onto each Ca site and the solution energies calculated in order to determine the degree of site selectivity, and if there is any change in the site selectivity with differing ionic radii. The cations Li^+, Na^+, K^+, Rb^+ and Cs^+ were chosen in order to give a spread of ionic radii both larger and smaller than that of Ca^{2+}. The results from these substitutions are displayed below in Table 1.

Table 1. Solution energies for substitution of Li^+, Na^+, K^+, Rb^+ and Cs^+ onto each Ca site of calcium chlorosilicate.

Dopant Cation	Ionic radii / Å	$E_{solution}$ / eV		
		Ca1 site	Ca2 site	Ca3 site
Li^+	0.92	0.989	1.954	2.120
Na^+	1.18	1.573	3.438	3.681
K^+	1.51	1.932	3.659	3.511
Rb^+	1.61	2.576	4.529	4.273
Cs^+	1.74	3.567	5.812	5.648

As can be seen, the Ca1 site is clearly favoured for the substitution of monovalent cations. This corresponds well with the understanding of the calcium chlorosilicate lattice, the Ca1 sites being the most weakly bonded, lying in between the lamellar layers of SiO_4^{4-} tetrahedra and so therefore the most easily substituted. In all cases there is an energy cost associated with the substitution, which becomes increasingly unfavourable as the ionic radii of the dopant cations increase. The Ca2 and Ca3 sites are much more strongly disfavoured, with minimal difference between the solution energies of the two sites. However, in all cases substitution of a monovalent cation is far more favourable than the creation of a Ca vacancy [5].

Charge-compensated Trivalent Substitution

The altervalent substitution of a trivalent cation onto a divalent Ca site can therefore be charge-compensated by the substitution of a monovalent charge-balancing species onto a second Ca site, as opposed to the creation of a highly unfavourable Ca vacancy. Such as substitution process can be described by the equation:

$$ACl + BCl_3 + 2Ca_{Ca(x)}^x \rightarrow A'_{Ca(x)} + B^{\bullet}_{Ca(x)} + 2CaCl_2 \qquad (2)$$

where A is the monovalent charge-balancing cation and B is the trivalent dopant cation.

The lanthanide cations Ho^{3+}, Gd^{3+}, Nd^{3+} and La^{3+} were substituted into the calcium chlorosilicate lattice in order to give a spread of ionic radii both larger and smaller than that of Ca^{2+}. Each were substituted in turn with Li^+, Na^+, K^+, Rb^+ or Cs^+ as the charge-balancing species in order to cover all possible permutations of cation pairings. The solution energy results for the lowest solution energy configuration of each pair of cations are shown in Table 2.

Table 2. Solution energies for lowest energy configuration of each pair of cations. [a] Equivalent solution energy for one lanthanide cation substituted onto the Ca2 site and charge compensated by a Ca vacancy on the Ca1 site [5].

Ca1 site dopant	Ionic radii / Å	$E_{solution}$ / eV			
		Ca2 = Ho	Ca2 = Gd	Ca2 = Nd	Ca2 = La
Li^+	0.92	2.595	2.840	3.577	4.208
Na^+	1.18	3.270	3.515	4.252	4.883
K^+	1.51	3.629	3.874	4.611	5.242
Rb^+	1.61	4.272	4.517	5.254	5.885
Cs^+	1.74	5.263	5.508	6.245	6.876
Vacancy[a]		3.711	3.956	4.693	5.324

In all cases, the lowest solution energy occurs when the trivalent cation is substituted onto the Ca2 site, allowing the monovalent charge-balancing species to be substituted onto its preferred Ca1 site, the reduced energy cost of its substitution there enough to compensate for forcing the trivalent cation onto the Ca2 site. In this respect, charge compensation via this method follows the trend observed when charge-balancing through vacancy formation, with the vacancy formed on the more favourable Ca1 site and the trivalent cation substituted onto the Ca2 site [5].

In all cases the solution energy increases with increasing ionic radius, the substitution process becoming more and more unfavourable with the increasing steric strain on the lattice. The equivalent solution energy for charge-balancing a single trivalent substitution with a vacancy is shown in Table 2 for comparison. Charge compensation via substitution of either of the smaller monovalent cations of Li^+ or Na^+ is clearly more preferable to the system due to the combination of the coulombic effects of only having to stabilise a net -1 charge on the defect as opposed to the net -2 charge of a vacancy and the lack of any significant steric stresses due to the substituted cation being similar or smaller in size than the host Ca^{2+}. Despite its larger size, charge-balancing with K^+ is also shown to be slightly favourable, although the difference in solution energy between it and charge compensation via vacancy formation is minimal, less than 0.1 eV in all cases.

Charge-balancing with the larger monovalent cations of Rb^+ or Cs^+ is less favourable. The increasing steric effects of substituting a much larger cation onto the Ca^{2+} host site, combined with those from the substitution of the lanthanide cation, now outweigh the energy savings of stabilising a net -1 charge over a net -2 charge, making charge compensation via vacancy formation the preferred option.

Charge-compensated multi-defect systems

Such a multi-defect system can be described by the equation:

$$ACl + MCl_2 + BCl_3 + 3Ca^x_{Ca(x)} \rightarrow A'_{Ca(x)} + M^x_{Ca(x)} + B^{\bullet}_{Ca(x)} + 3CaCl_2 \qquad (3)$$

where A is the monovalent charge-balancing cation, M is the divalent dopant cation and B the trivalent dopant cation. The solution energy results for the lowest solution energy configuration of each group of cations are shown in Table 3.

Table 3. Solution energies for lowest energy configuration of cations.

Ca1 site	Ca2 site	Ca3 site	$E_{solution}$ / eV
A'	Fe^{2+}	La^{3+}	7.2 - 9.8
A'	Fe^{2+}	Nd^{3+}	6.6 - 9.2
A'	Gd^{3+}	Fe^{2+}	5.9 - 8.5
A'	Ho^{3+}	Fe^{2+}	5.6 - 8.3
A'	Mn^{2+}	La^{3+}	7.6 - 10.3
A'	Mn^{2+}	Nd^{3+}	7.0 - 9.7
A'	Gd^{3+}	Mn^{2+}	6.4 - 9.1
A'	Ho^{3+}	Mn^{2+}	6.1 - 8.8
Sr^{2+}	La^{3+}	A'	6.6 - 9.9
Sr^{2+}	Nd^{3+}	A'	6.0 - 9.2
Sr^{2+}	Gd^{3+}	A'	5.2 - 8.5
Sr^{2+}	Ho^{3+}	A'	5.0 - 8.2

Initially the trend seen is identical to that observed for charge compensation via vacancy formation. When the smallest of the divalent cations used in this work, Fe^{2+}, is paired with either of the larger trivalent cations (i.e. La^{3+} or Nd^{3+}) then, regardless of the charge-balancing species used, the monovalent cation will always be substituted onto its preferred Ca1 site, forcing the di- and trivalent cations on to their second preference sites of Ca2 and Ca3 respectively. However, as the ionic radii of the trivalent cations decreases the difference between the solution energies of the three Ca sites increases [5]. When paired with either of the smaller trivalent cations (i.e. Gd^{3+} or Ho^{3+}), whilst the monovalent cation remains on the Ca1 site, Fe^{2+} is now substituted on its least favoured Ca3, with Gd^{3+} or Ho^{3+} now substituted onto their preferred Ca2 site, the decrease in solution energy more than enough to compensate for the extra energy penalty incurred by forcing the Fe^{2+} cation onto its least favoured Ca3 site.

When Sr^{2+} is the divalent cation used then, regardless of the tri- or monovalent cations substituted with it, Sr^{2+} will always be substituted onto its preferred Ca1 site, the trivalent cation onto its preferred Ca2 site and the monovalent cation onto the Ca3 site. This is due entirely to the large difference in solution energies of substituting the Sr^{2+} onto the Ca1 and Ca2 sites (1.65 eV) [5], the energy saving of having both di- and trivalent cations on their preferred sites more than enough to compensate for the energy penalty incurred by forcing the monovalent cation onto the Ca3 site.

For multi-defect substitutions involving the smaller divalent cations of Fe^{2+} or Mn^{2+}, charge-compensation via substitution of either of the smaller monovalent cations of Li^+ or Na^+ is preferred to vacancy formation on a coulombic basis. K^+ is also shown to be slightly favourable, although once again the difference in solution energy between it and charge compensation via vacancy formation is less than 0.1 eV. Charge-balancing with the larger monovalent cations of Rb^+ or Cs^+ is less favourable, again due to the increasing steric effects of substituting a much larger cation onto the Ca^{2+} host site.

For multi-defect substitutions involving Sr^{2+} as the divalent cation however, only charge-balancing with Li^+ is more favourable than via vacancy formation. With the now much larger Sr^{2+} substituted onto the Ca1 site, very little additional steric strain from the other substitutions is needed to outweigh the purely coulombic benefits of stabilising the net -1 charge of the monovalent defect over the net -2 charge of the vacancy.

CONCLUSIONS

In the substitution of isolated monovalent cations, a clear preference for the Ca1 site is displayed. The Ca2 and Ca3 sites are equally disfavoured, with minimal difference in solution energies between them, in common with the case for Ca vacancy formation. Substitution of a monovalent cation into the calcium chlorosilicate lattice becomes less favourable with the increasing ionic radius of the dopant cation, as the steric stresses associated with substitution of larger cations than the Ca^{2+} host increase.

When charge compensating the substitution of a trivalent cation by co-substitution of a monovalent cation, a clear site selectivity is identified, the monovalent cation preferring the Ca1 site with the trivalent dopant substituting onto the Ca2 site. The solution energy of the substitution process increases with the increasing ionic radii as the steric stresses associated with substitution of larger cations than the Ca^{2+} host increase. As such, only charge-balancing using Li^+, Na^+ or K^+ is more favourable than via formation of a Ca vacancy.

In multi-defect systems featuring the substitution of both di- and trivalent dopant cations into the calcium chlorosilicate lattice, it is the ionic radius of the divalent dopant that governs the site selectivity of the charge-balancing monovalent species. With the smaller divalent cations, charge-balancing using Li^+, Na^+ or K^+ is again more favourable than via formation of a Ca vacancy. However, for the larger divalent cations, the additional steric strain placed upon the lattice means that only charge-balancing with Li^+ is favourable to charge compensation via vacancy formation.

REFERENCES

1. G. Leturcq, A. Grandjean, D. Rigaud, P. Perouty, M. Charlot, *J. Nucl. Mater.,* **347**, 1, (2005).
2. E. N. Treushnikov, V. V. Ilyukhin, N. V. Belov, *Doklady Akademii Nauk SSSR,* **193**, 1048 (1970).
3. R. Czaya, G. Bissert, *Kristall und Technik,* **5**, 9 (1970).
4. Q. Shen, M. Lu, H. Lin, C. Li, *Adv. Mater. Res.,* **216**, 16, (2011).
5. M. R. Gilbert, *J. Phys. Chem. Solids,* **75**, 1004, (2014).
6. J. D. Gale, J. Chem. Soc. Faraday Trans., **93**, 629, (1997).
7. C. R. A. Catlow, W. C. Mackrodt, *Computer Simulation of Solids*, (Springer, Berlin 1982).
8. D. F. Shanno, *Math. Comp.,* **24**, 647, (1970).
9. P. P. Ewald, *Ann. Phys.,* **64**, 253, (1921).
10. B. G. Dick, A. W. Overhauser, *Phys. Rev.,* **112**, 90, (1958).
11. N. F. Mott, M. J. Littleton, *Trans. Faraday Soc.,* **34**, 485, (1938).
12. S. T. Murphy, H. Lu, R. W. Grimes, *J. Phys. Chem. Solids,* **71**, 735, (2010).
13. M. Dove, *Am. Mineral.,* **74**, 774, (1989).
14. J. A. L. Rabone, N. H. de Leeuw, *J. Comput. Chem.,* **27**, 266, (2006).

Mater. Res. Soc. Symp. Proc. Vol. 1744 © 2015 Materials Research Society
DOI: 10.1557/opl.2015.396

Effect of Charge-balancing Species on Sm^{3+} Incorporation in Calcium Vanadinite

M. R. Gilbert
AWE, Aldermaston, Reading, RG7 4PR, UK.

ABSTRACT

Apatites are often seen as good potential candidates for the immobilization of halide-rich wastes. In particular, phosphate apatites have received much attention in recent years, however, their synthesis often produces complicated multi-phase systems, with a number of secondary phases forming [1.2]. Calcium vanadinite ($Ca_5(VO_4)_3Cl$) demonstrates a much simpler phase system, with only a single $Ca_2V_2O_7$ secondary phase which can easily be retarded by the addition of excess $CaCl_2$. However, when doping with $SmCl_3$ (as an inactive analogue for $AnCl_3$) the Sm forms a wakefieldite ($SmVO_4$) phase rather than being immobilized within the vanadinite, a result of having to form an energetically unfavourable Ca vacancy in order for the lattice to remain neutral overall. It has been postulated that charge-balancing the lattice via co-substitution of a monovalent cation will be less disfavoured and therefore help stabilise formation of a $(Ca_{5-2x}Sm_xA_x)(VO_4)_3Cl$ solid solution (A = monovalent cation). This has been investigated using a combined modelling and experimental approach. Static lattice calculations performed using Li^+, Na^+ and K^+ as charge-balancing species have shown the energy cost to be less than half that of charge-balancing via formation of a Ca vacancy. As a result, solid state synthesis of $(Ca_{5-2x}Sm_xLi_x)(VO_4)_3Cl$, $(Ca_{5-2x}Sm_xNa_x)(VO_4)_3Cl$ and $(Ca_{5-2x}Sm_xK_x)(VO_4)_3Cl$ solid solutions have been trialled, and analysis of the resulting products has shown a significant reduction in both the $SmVO_4$ and $Ca_2V_2O_7$ secondary phases across all dopant levels.

INTRODUCTION

Calcium vanadinite ($Ca_5(VO_4)_3Cl$) is a chlorapatite isomorph, crystallising in the hexagonal $P6_3/m$ system with the Cl⁻ anions lying in the [00z] channels. Vanadinites have been viewed as prospective waste-forms for a number of years, with lead iodo-vanadinite ($Pb_5(VO_4)_3I$) proposed as a potential host for [129]I, the iodine able to be immobilized within the channels of the z-axis [3,4]. Synthesis of single-phase $Ca_5(VO_4)_3Cl$ is documented in the literature using a two-step process, firstly reacting $CaCO_3$ and V_2O_5 in a 3:1 molar ratio at 1100 °C for 10 h to form $Ca_3(VO_4)_2$, then reacting this with $CaCl_2$ in a 3:1 molar ratio at 900 °C repeatedly for 1 h [5,6]. These processing conditions are too extreme to meet AWE's requirements (limit of 800 °C), so a low temperature, single-step synthesis has been successfully developed, allowing the production of near single-phase calcium vanadinite at 750 °C in 5 h [7].

However, attempted incorporation of $SmCl_3$ (as an inactive analogue for $PuCl_3$ and $AmCl_3$) into the calcium vanadinite did not yield a single-phase solid solution. Whilst the Cl was successfully immobilized within the vanadinite, the Sm formed a separate $Sm(VO_4)$ tetragonal wakefieldite phase [7]. In trying to substitute a trivalent cation onto a divalent site in the lattice, vacancies must also be created in order for the lattice to remain charge neutral. It is hypothesised that the creation of these vacancies presents an insurmountable barrier to the substitution process, resulting in the formation of the $Sm(VO_4)$ wakefieldite phase instead of a $(Ca_{(5-1.5x)}Sm_x)(VO_4)_3Cl$ solid solution.

To better understand this process, different charge compensation mechanisms for the calcium vanadinite have been investigated using lattice dynamics calculations to identify the most energetically favourable mechanism. These have then been employed in the synthesis of Sm-doped calcium vanadinite, in an attempt to form a single-phase solid solution as opposed to a separated, two-phase system.

EXPERIMENT

Sm-doped calcium vanadinite was produced via a solid state synthesis (SSS) reaction as shown in Equation 1, where A is a monovalent cation.

$$9CaO + 3V_2O_5 + (1-2x)CaCl_2 + xSmCl_3 + xACl \rightarrow 2(Ca_{5-2x}Sm_xA_x)(VO_4)_3Cl \qquad (1)$$

Stoichiometric quantities of CaO, $CaCl_2$, V_2O_5, $SmCl_3$ and either LiCl, NaCl or KCl, together with a 1 mol excess of $CaCl_2$ to retard the formation of the $Ca_2V_2O_7$ secondary phase, were placed into a Nalgene mill pot together with yttria-stabilised zirconia (YSZ) milling media and dry milled overnight. The resulting powder mix was passed through a 250 μm sieve mesh, placed in an alumina crucible, heated to 300 °C at 3.4 °C/min and held for 1 h to drive off any absorbed moisture. This dried powder was then heated at 3.4 °C/min to 800 °C and fired in air for 5 h in order to assess the effect of calcination temperature on the final product.

The calcined plugs were broken up, passed through a 250 μm sieve mesh and characterised by X-ray diffraction (XRD). Powder XRD was carried out using a Bruker D8 Advance diffractometer operating in Bragg-Brantano flat plane geometry using Cu $K_{\alpha 1}$ radiation (λ = 1.54056 Å). Diffraction patterns were measured over a 2θ range of 10 – 90 ° using a step size of 0.025 ° and a collection time of 3.4 s per step. The phases present were identified by pattern matching using the Bruker Eva software and PDF database. Refinement of the phase assemblage and unit cell parameters was carried out using the GSAS suite of software [8].

RESULTS

$(Ca_{5-2x}Sm_xLi_x)(VO_4)_3Cl$

Figure 1 shows the XRD patterns of $(Ca_{5-2x}Sm_xLi_x)(VO_4)_3Cl$ (where x = 0.1, 0.2 and 0.3) synthesised at 800 °C. The phase assemblage of the products is shown in Table 1.

Table 1. Phase assemblage of products arising from synthesis of $(Ca_{5-2x}Sm_xLi_x)(VO_4)_3Cl$.

x	$Ca_5(VO_4)_3Cl$ / wt. %	$Ca_2V_2O_7$ / wt. %	$Sm(VO_4)$ / wt. %	$Ca_2(VO_4)Cl$ / wt. %	R_{wp}
0.1	89.1	7.5	1.2	2.3	4.72
0.2	87.1	6.4	3.2	3.2	4.70
0.3	49.7	11.8	6.6	31.9	2.97

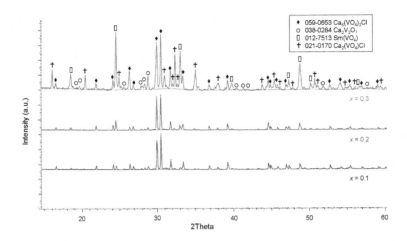

Figure 1. XRD patterns of $(Ca_{5-2x}Sm_xLi_x)(VO_4)_3Cl$ fired in air at 800 °C for 5 h.

As can be seen in the XRD patterns, large amounts of secondary phases are still present in the Li^+ compensated system. It should be noted that the levels of separated $Sm(VO_4)$ wakefieldite phase are lower than in the vacancy compensated system (5, 8 and 10 wt. % for $x = 0.1$, 0.2 and 0.3 respectively) [7], indicating increased levels of incorporation within the vanadinite lattice relative to that system. This does not preclude the potential incorporation in the other vanadium-containing phases as it is not possible to determine this conclusively by XRD at such low levels. However, instead of a concomitant rise in the percentage of the $Ca_5(VO_4)_3Cl$ relative to the vacancy compensated system, it is the levels of the $Ca_2V_2O_7$ and $Ca_2(VO_4)Cl$ secondary phases that increase. Particularly striking is the appearance of the $Ca_2(VO_4)Cl$ spodiosite isomorph, which is absent from the vacancy compensated system when fired at temperatures $\geq$ 750 °C. The highest substitution level, $x = 0.3$, produces a completely different distribution of phases, with a substantial increase in the amounts of $Ca_2V_2O_7$ and $Ca_2(VO_4)Cl$ phases, such that the target $Ca_5(VO_4)_3Cl$ phase is now < 50 wt. %.

$(Ca_{5-2x}Sm_xNa_x)(VO_4)_3Cl$

Figure 2 shows the XRD patterns of $(Ca_{5-2x}Sm_xNa_x)(VO_4)_3Cl$ (where $x = 0.1$, 0.2 and 0.3) synthesised at 800 °C. The phase assemblage of the products is shown in Table 2.

Table 2. Phase assemblage of products arising from synthesis of $(Ca_{5-2x}Sm_xNa_x)(VO_4)_3Cl$.

x	$Ca_5(VO_4)_3Cl$ / wt. %	$Ca_2V_2O_7$ / wt. %	$Sm(VO_4)$ / wt. %	R_{wp}
0.1	92.8	5.9	1.3	4.56
0.2	97.7	0.0	2.3	5.25
0.3	94.3	0.0	5.8	4.01

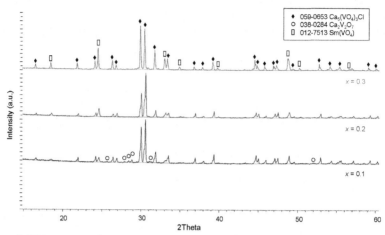

Figure 2. XRD patterns of $(Ca_{5-2x}Sm_xNa_x)(VO_4)_3Cl$ fired in air at 800 °C for 5 h.

XRD of the Na^+ compensated system shows a much simpler phase assemblage than that charge-balanced with Li^+. At the initial substitution level of $x = 0.1$, the $Ca_2V_2O_7$ secondary phase is present in approximately the same amount as in the vacancy compensated system (7.0 wt. % compared to 7.8 wt. % respectively) [7], but as the substitution level increases this phase disappears from the product. The levels of separated $Sm(VO_4)$ wakefieldite phase are again lower than in the vacancy compensated system, indicating increased levels of incorporation within the vanadinite lattice, and unlike the Li^+ compensated system, the percentage of $Ca_5(VO_4)_3Cl$ has increased, > 90 wt. % in all cases compared to 85 - 87 wt. % in the vacancy compensated system [7].

$(Ca_{5-2x}Sm_xK_x)(VO_4)_3Cl$

Figure 3 shows the XRD patterns of $(Ca_{5-2x}Sm_xK_x)(VO_4)_3Cl$ (where $x = 0.1, 0.2$ and 0.3) synthesised at 800 °C. The phase assemblage of the products is shown in Table 3.

Table 3. Phase assemblage of products arising from synthesis of $(Ca_{5-2x}Sm_xK_x)(VO_4)_3Cl$.

x	$Ca_5(VO_4)_3Cl$ / wt. %	$Ca_2V_2O_7$ / wt. %	$Sm(VO_4)$ / wt. %	KVO_3 / wt. %	R_{wp}
0.1	97.4	1.2	1.5	0.0	5.60
0.2	95.5	0.9	3.5	0.0	4.40
0.3	93.3	0.0	5.7	1.0	4.53

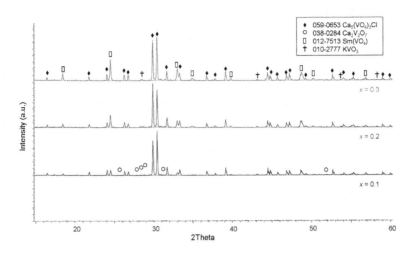

Figure 3. XRD patterns of $(Ca_{5-2x}Sm_xK_x)(VO_4)_3Cl$ fired in air at 800 °C for 5 h.

The results of charge-compensation using K^+ are very similar to those achieved by the Na^+ compensated system, with near identical amount of $Sm(VO_4)$ formed for each substitution level. As with the Na^+ compensated system, no $Ca_2(VO_4)Cl$ secondary phase is formed. At the lowest substitution level, $x = 0.1$, the percentage of $Ca_2V_2O_7$ present is much lower than in the Na^+ compensated system, however, trace amounts are still present at $x = 0.2$. At the highest substitution level, $x = 0.3$, although this $Ca_2V_2O_7$ secondary phase is now absent, in the K^+ compensated system it is replaced by a new KVO_3 phase, resulting in a corresponding drop in the level of target $Ca_5(VO_4)_3Cl$ formed.

DISCUSSION AND CONCLUSIONS

Whilst the $(Ca_{5-2x}Sm_xLi_x)(VO_4)_3Cl$ system does produce lower levels of separated $Sm(VO_4)$ wakefieldite phase than the vacancy compensated approach, there are also greatly increased levels of $Ca_2V_2O_7$ and $Ca_2(VO_4)Cl$ secondary phases. Particularly of note is the $Ca_2(VO_4)Cl$ spodiosite isomorph, which is absent from the vacancy compensated system, but here becomes almost the major phase at the highest substitution level. This is due to the contraction of the average radius of the A-site in the calcium vanadinite lattice, the site onto which the Li^+ and Sm^{3+} are substituted. The Li^+ cation is the only charge-balancing species used that is smaller then the original Ca^{2+} (ionic radius of 0.76 Å as opposed to 1.0 Å for Ca^{2+}). As a result, its substitution will cause the average A-site radius to contract, as shown in Figure 4.

The average A-site radius will continue to decrease with increasing levels of Li^+ substitution and as a result a transformation to the smaller $Ca_2(VO_4)Cl$ phase (unit cell volume = 498.01 Å^3 compared to 606.07 Å^3 for $Ca_5(VO_4)_3Cl$) is observed.

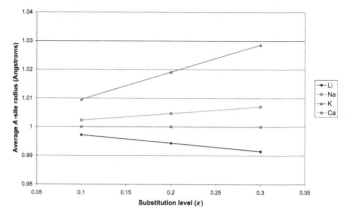

Figure 4. Graph of change in average radius of the A-site in the calcium vanadinite lattice with increasing substitution level. Ca is shown for reference.

In contrast, the Na^+ and K^+ cations are both larger than the original Ca^{2+} (ionic radii of 1.02 and 1.38 Å respectively) and will therefore cause an expansion of the average A-site radius. As a result, there is no transition to the $Ca_2(VO_4)Cl$ phase observed in either the $(Ca_{5-2x}Sm_xNa_x)(VO_4)_3Cl$ or $(Ca_{5-2x}Sm_xK_x)(VO_4)_3Cl$ system. The two systems are very similar in behaviour, both producing near identical quantities of separated $Sm(VO_4)$ wakefieldite. However, with the larger ionic radius, the rate of increase in the average A-site radius of the $(Ca_{5-2x}Sm_xK_x)(VO_4)_3Cl$ system is obviously greater, such that at $x = 0.3$ a separated KVO_3 phase is now present, the calcium vanadinite lattice no longer able to accommodate any more K^+.

In the $(Ca_{5-2x}Sm_xNa_x)(VO_4)_3Cl$ system, the difference between the ionic radii of both the Na^+ and Sm^{3+} and the original Ca^{2+} is minimal (0.02 Å for the Na^+ and 0.1 Å for the Sm^{3+}). As a result, the rate of increase in the average A-site radius is very slow and thus minimises the steric stresses from the co-substitution of Na^+ and Sm^{3+} into the vanadinite lattice, allowing it to remain stable at the higher substitution levels.

REFERENCES

1.   S. K. Fong, I. W. Donald, B. L. Metcalfe, *J. Alloys Comp.*, **444/445,** 424 (2007).
2.   B. L. Metcalfe, I. W. Donald, S. K. Fong, L. A. Gerrard, D. M. Strachan, R. D. Scheele, *Mat. Res. Soc. Symp. Proc.*, **985,** 157 (2007).
3.   F. Audubert, J. Carpena, J. L. Lacout, F. Tetard, *Solid Sate Ionics*, **95,** 113 (1997).
4.   M. C. Stennett, I. J. Pinnock, N. C. Hyatt, *J. Nucl. Mater.*, **414,** 352 (2011).
5.   H. Kreidler, *Am. Mineral.*, **55,** 180 (1970)
6.   H. P. Beck, H. Douiheche, R. Haberkorn, H. Kohlmann, *Solid State Sci.*, **8,** 64 (2006)
7.   M. R. Gilbert, *Mat. Res. Soc. Symp. Proc.*, **1665,** 319, (2014).
8.   H. M. Rietveld, *Acta Cryst.*, **22,** 151 (1967).

Corrosion Behavior of Materials

Mater. Res. Soc. Symp. Proc. Vol. 1744 © 2015 Materials Research Society
DOI: 10.1557/opl.2015.317

Key Phenomena Governing HLW Glass Behavior in the French Deep Geological Disposal

Stéphan Schumacher[1], Christelle Martin[1], Yannick Linard[1], Frédéric Angeli[2], Delphine Neff[3], Abdesselam Abdelouas[4] and Xavier Crozes[5]

[1]ANDRA, 1/7, rue Jean Monnet, Parc de la Croix-Blanche, 92298 Châtenay-Malabry, France
[2]CEA, DEN, Laboratoire du Comportement à Long Terme, BP 17171, 30207 Bagnols/Cèze, France.
[3]SIS2M/LAPA, CEA Saclay, 91191 Gif sur Yvette Cedex, France
[4]SUBATECH – EMN-CNRS/IN2P3-Université de Nantes, 4 rue Alfred Kastler – La Chantrerie, B.P. 20722, 44307 Nantes cedex 03, France
[5]EDF R&D, Moret Sur Loing, France

ABSTRACT

According to the Planning Act of 28th June 2006, Andra is in charge of ensuring the sustainable management of all radioactive waste generated in France, especially the high-level and long-lived vitrified waste produced from spent fuel recycling.

Since 2006, all the studies and research related to the components of HLW cells have been incorporated into a broader R&D program which aims at characterizing and modeling (i) the glass matrix dissolution, (ii) the corrosion of the overpack and the lining, and (iii) the claystone evolution in the near field, considering all the interactions between these surrounding materials. This program, coordinated by Andra, has involved up to eighteen laboratories.

After closure of disposal cells and overpack failure, glass alteration is expected to begin in partially saturated conditions due to hydrogen production resulting from carbon steel corrosion in anoxic conditions. Therefore, the glass should at least partially be hydrated by water vapor during thousands of years until complete saturation. A part of the studies aimed to determine the glass behavior in such conditions, the influence of the main parameters (temperature, relative humidity) and consequences of vapor hydration on subsequent radionuclides release by water leaching.

In addition, the major part of the work focused on the influence of the environment on glass alteration. The effect of clay pore water on glass alteration rates (initial rate, rate drop and residual rate) was determined and particularly that of pH and magnesium. The nature of steel corrosion products and their interactions with glass alteration were also investigated. All these studies relied on experiments in surface laboratories, in Andra's underground laboratory, together with natural or archeological analogs and modeling studies.

INTRODUCTION

According to the Planning Act of 28th June 2006, Andra is in charge of ensuring the sustainable management of all radioactive waste generated in France, especially the high-level and long-lived vitrified waste produced from spent fuel recycling.

During the spent fuel recycling, the fission products and minor actinides are separated and incorporated into borosilicate glass. Then they are poured into a container which is made of refractory stainless steel and referred to as the primary container.

The primary container is placed in a disposal container made of 55 mm thick carbon steel shell and a lid (figure 1) before being disposed of in cells drilled in Callovo-Oxfordian claystone. The main function of this overpack is to prevent water from reaching the vitrified HLW at least during the thermal phase (>50 °C) which lasts for about a thousand years [1].

The HLW disposal cells are horizontal micro-tunnels with a diameter of 0.7 m. They are lined with a metal tube into which the waste packages can slide (figure 1). The disposal container must contribute to dissipate the heat coming from the packages. As they are particularly exothermic, the packages are separated by spacing buffers similar to the disposal packages so as to comply with the thermal constraints of the repository (the temperature on the outer face of the package should be less than 100 °C) and to optimise the management of the underground site [1].

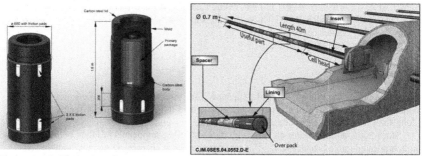

Figure 1. Vitrified waste disposal package and HLW disposal cell, configuration after closure of the cell [1]

Since 2006, all the studies and research related to the components of HLW disposal cells have been incorporated into a broader R&D program which aims at characterizing and modeling (i) the glass matrix dissolution, (ii) the corrosion of the overpack and the lining, and (iii) the claystone evolution in the near field. This program has taken into account all the interactions between these materials and the different situations likely to occur in repository. This paper is a review of the studies on glass alteration carried out within this program.

EXPECTED SITUATIONS IN REPOSITORY CONDITIONS

The excavation of the cell and the operating period lead to air penetration and consequently partial desiccation and oxidation of the claystone. After the closing of geological disposal cells, the HL waste packages are then expected to be exposed to an environment that will evolve with time from a hot and humid atmosphere containing little air to a clayey groundwater in anoxic conditions at the temperature of the geological medium (figure 2).

Owing to the radioactivity of the fission products (mainly of radionuclides ^{137}Cs and ^{90}Sr), the vitrified waste packages generate high amount of heat. After rising up to 100 °C for a few years, the temperature of the packages gradually decreases due to radioactive decay. It will take 500 to 1 000 years for the temperature to reach 50 °C depending on the radioactive inventory.

During this so-called "thermal phase," the overpack protects the nuclear glass from any water ingress from the host rock. The vitrified waste is first in a dry environment (Situation 1, figure 2). Once the overpack is no longer watertight the glass comes into contact with water. In these conditions, only three phenomena may modify its physico-chemical characteristics:
- the crystallization,
- the radiation damage induced by the radioactive decay of actinides and fission products immobilized in the borosilicate glass matrix,
- and the mechanical constraints induced by the thermal gradients during the vitrification and cooling processes.

According to the design of the cells (small clearance volume) and the repository management methods, the oxygen trapped in the cell and in the claystone damaged during excavation will be rapidly consumed by corrosion (mainly the outer surface of the lining and the cell head) or by other physical-chemical processes, such as the oxidation of the claystone. Then anoxic conditions will prevail in parallel with the resaturation process. The interface between the clay and the lining and then the gap between the lining and the overpack will gradually be filled with groundwater. The corrosion of metal components will therefore occur in clay water or in contact with the claystone from the excavation damaged zone (Situation 2, figure 2).

Due to corrosion and mechanical constraints, the overpack and the primary container will finally breach. As a result the glass might be in contact with groundwater. However, the hydrogen production induced by the anoxic corrosion of metallic components is likely to prevent a fast resaturation of the voids within the waste package [1] (Situation 3, figure 2). As a consequence glass vapor hydration may occur before the groundwater has completely infiltrated the container (Situation 4, figure 2).

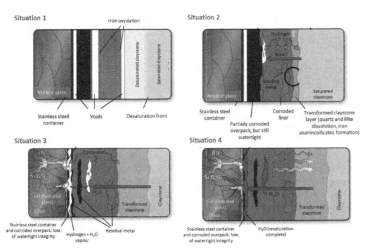

Figure 2. Evolution of the different interfaces in the HL disposal cells during the different situations expected in repository conditions

R&D PROGRAM AND MAIN RESULTS

The objective of the research carried out since 2006 is to determine the physical-chemical behaviour of the nuclear glass in the different expected situations (figure 2) and its interactions with the surrounding materials (metallic components, claystone).

Glass vapor hydration

Canister breaching marks the beginning of glass alteration. However hydrogen production resulting from carbon steel corrosion will lead to unsaturated conditions. As a result glass alteration will occur in "atmospheric conditions" with a relative humidity (RH) close to 100 % and a temperature below 50 °C. In these conditions, a thin layer of water adsorbs on the glass surface and its thickness depends on the type of glass, the alteration time and the relative humidity. The amount of water adsorbed on a SON68 glass (inactive surrogate of the French R7/T7 HLW glass) monolith altered for 99 days at 92 % RH and 175 °C is greater than that adsorbed on a pristine glass (figure 3). The thickness of the layer increases drastically when relative humidity reaches 80-90 %.

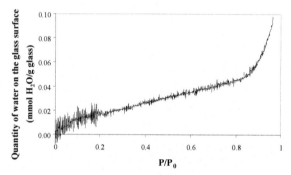

Figure 3. Quantity of water adsorbed on a SON68 glass altered for 99 days at 92 % RH and 175 °C (from [2])

Glass alteration rate increases with relative humidity (from 92 % RH up to 99.9% RH) and with temperature (from 50 °C up to 200 °C) but decreases over time after several hundred days (figure 4). The activation energy ranges from 43 to 47 kJ.mol^{-1} between 125 °C and 200 °C at 92 % RH [2,3] but recent studies [4] have showed a lower activation energy between 35 °C and 125 °C (95 % RH) in accordance with Gong studies [5]. The hydration reaction rate at 90 °C is one order of magnitude higher than the residual rate in liquid water at the same temperature. The difference is greater when the temperature is lower but the hydration reaction rate keeps decreasing [4].

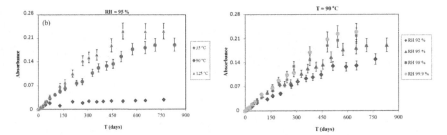

Figure 4. Evolution with time of the absorbance of the glass as a function of RH (on the left) and of temperature (on the right) (from [4]). The relative humidity of these experiments was maintained by placing solutions of different weight percent NaCl in the autoclave.

The secondary phases have been characterized using TEM, SEM and Raman spectroscopy. They mainly depend on temperature and alteration time. The major precipitates are calcite, apatite (from 90 °C), powellite (from 90 °C), tobermorite (from 125 °C) and analcime (from 125 °C). Some of these precipitates trap silicon and therefore increase glass alteration rate [2]. Cross sectional analysis has also shown amorphous Si/Al precipitates with minor amount of Fe and Ni [4].

The mechanisms behind glass vapor hydration are still unknown but the presence of C-H-S and zeolites suggests an increase of pH of the water thin film on the glass surface. Furthermore, some experiments carried out with different gases in order to control the pH have shown that glass hydration rate increases with increasing pH [6].

Leaching of pre-hydrated glass leads to a quick increase of pH and a higher release of boron and lithium. The release increases with increasing RH and therefore with glass gel layer thickness (figure 5). The release decreases with time which suggests that the hydrated gel layer is more susceptible to leaching than a pristine glass. From a safety assessment point of view, these results imply that the radionuclides inventory localized in the pre-hydrated gel corresponds to a labile fraction.

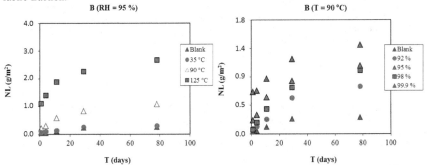

Figure 5. Normalized mass loss of boron as a function of time for samples hydrated at different temperatures at 95 % relative humidity (on the left) or at different relative humidity at 90 °C (on the right) and then leached in clayey water at 50 °C for 78 days (from [4]).

Glass alteration in clay-equilibrated groundwater

Once hydrogen production decreases enough for the groundwater to fill the canister, glass alteration will occur in saturated conditions in clay-equilibrated water. In such conditions, solution chemistry will have an important effect on all the stages of glass alteration (forward rate, rate drop regime and residual rate).

The forward rate in clay-equilibrated groundwater is about 5.5 times greater than in deionized water but the activation energy is similar (77±3.2 kJ/mol, figure 6). Two phenomena account for this increase: the ionic strength of the clay-equilibrated groundwater and the alkali-metal and alkali-earth concentration (mainly calcium ions). The underlying mechanism relies on the speciation of glass surface sites induced by these elements that makes the silicon–oxygen bonds more readily hydrolysable [7].

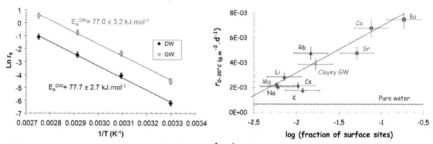

Figure 6. (left) forward dissolution rate, r_0 (g·m^{-2}·d^{-1}), of SON68 glass versus temperature measured in deionized water (DW) and clay-equilibrated groundwater (GW) (from [7])– (right) Forward dissolution rate, r_0, of SON68 glass at 20 °C in presence of different cations at a ionic strength of 0.1 mol·L^{-1} and in clay-equilibrated groundwater (GW) versus the fraction of surface metal sites for alkali metal and alkaline earth cations (from [7])

Clay-equilibrated water delays the rate drop leading to a greater amount of altered glass than in deionized water. Moreover, not only are the glass alteration tracers anti-correlated with the magnesium concentration but the pH of the leachate is also lower than that resulting from dissolution in deionized water (figure 7). Structural analysis by SEM-EDS of SON68 glass surface leached in clay-equilibrated water at 90 °C show phyllosilicate-type phases with spherical morphology and a layer of uniform thickness around the glass grains. Chemical analyses by EPMA reveal that the external layer contains a large quantity of magnesium that corresponds to a mixture of gel, phyllosilicates and magnesium silicates, while the phyllosilicates comprise mainly silicon and magnesium. Three magnesium silicates were identified with X-ray diffraction: sepiolite (Si/Mg=1.5), hectorite (Si/Mg=1.33) and saponite (Si/Mg=1.22) [8]. Modeling of the experiments with the GRAAL model [9] shows that a Si/Mg ratio equal to 1.5 leads to the best fit of the pH and the boron concentration. It also shows that the precipitation of secondary magnesium silicate phases leads to silicon consumption and subsequently to a higher leach rate and a release of H$^+$ ions inducing a drop in pH [8]. When all the magnesium in solution has been consumed and the pH has again increased to a value close to that of pristine clay-equilibrated water, the glass leaching rate drops to the same low values observed in pure water [8]. Glass leaching in groundwater with a low renewal ensuring a constant magnesium source leads to an increase of the residual rate by a factor close to 5 due to the pH decrease and the precipitation of the magnesium silicate secondary phases. It is essential to note that the precipitation of these phases only occurs if the pH is greater than 7 at 90 °C (greater than 8.3-8.4 at 50 °C). In repository conditions, reactive transport modeling has to be performed to assess the magnesium availability and the pH near the vitrified waste.

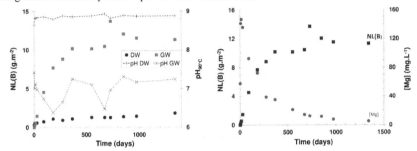

Figure 7. Normalized mass loss of boron and pH as a function of time for samples leached in deionized water (DW) and clay-equilibrated groundwater (GW) (on the left)–Normalized mass loss of boron and magnesium concentration as a function of time in clay-equilibrated water (on the right) (from [8])

Alteration in presence of environment materials

The water which is expected to leach the glass is presumably at equilibrium with the nearfield materials, including iron and/or its corrosion products and the surrounding Callovo-Oxfordian clay. The overpack corrodes first in groundwater and the resulting corrosion products (CP) are characteristic of a corrosion in clayey environment. As the overpack breaks due to mechanical constraints, part of it remains in the system and its corrosion is then concomitant to glass alteration. The nature of CP is likely to be influenced by the solution chemistry at the interface between the glass and the overpack and especially by the elements released by the glass alteration. Needless to say it is essential to determine how these nearfield materials (including corrosion products) can influence the alteration kinetics of nuclear glass in repository conditions, mainly through interactions with the silica released by the glass dissolution. More specifically some processes such as silica sorption and precipitation of silicate phases favor glass dissolution by delaying the saturation of the aqueous solution necessary for the formation of a protective gel [9,10].

Because of the predominantly reducing conditions, the corrosion of the steel overpack leads to the precipitation of Fe-oxides or carbonates such as magnetite, siderite and chukanovite. The corrosion of carbon steel in the conditions expected in the geological repository has been studied through long-term corrosion tests both in surface laboratories and in the Meuse/Haute-Marne underground laboratory in Bure, France [11-13]. The characterization of archaeological artefacts, such as nails buried for several centuries in an anoxic carbonated environment on a 16th-century steelmaking site, was also a means of supporting the experimental results on a larger time scale [14-17].

An important R&D program has been developed including several experiments of glass alteration in synthetic groundwater at 50 °C in presence of near-field materials in different configurations : (i) in homogeneous systems in which mixed glass and corrosion products CP (mainly magnetite) or iron powders are introduced in a reactor, (ii) in diffusion cells in which materials are separated by a membrane or a stainless steel filter, (iii) in heterogeneous systems consisting of piles of successive layers of materials (glass, CP, iron, claystone).

Glass / corrosion products interactions

Previous studies have already highlighted that silicon released from the glass is subjected to sorption on the corrosion products [18-22]. However, Philippini et al. [20] who studied silicon sorption on different commercial CP showed that sorption may have only a short-term influence compared to the lifetime of nuclear glass expected in repository conditions, limited by the sorption capacity of corrosion products.

Recent experimental results in homogeneous systems [23] have confirmed that the altered fraction deeply depends on the amount of CP added to the system. Godon et al. [23] modeled their experimental data with the GRAAL model [9] and a model of silicon sorption onto magnetite. Both these models were implemented in the geochemical code Hytec [24]. Godon et al. [23] pointed out that silicon sorption allows to simulate the glass alteration on the first ten days but doesn't allow to reproduce the longer-term trend. It seems necessary to consider another process, such as precipitation of an iron silicate phase, even if ongoing studies have not allowed to identify precisely the neoformed phase in the system so far (figure 8).

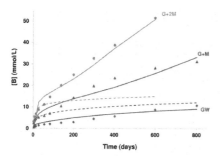

Figure 8. Comparison between experimental data results (diamonds) and modeling either with silicon sorption only (dashed lines) or with Si sorption + precipitation of an iron silicate phase (solid lines) obtained for the alteration of glass powder in presence of magnetite in synthetic groundwater at 50 °C (G+M: 10 g of glass + 10 g of magnetite; G+2M: 10 g of glass + 20 g of magnetite). The whole is compared with the alteration of only glass in synthetic groundwater (Blank) (from [23]).

Experiments carried out in reactor cells through successive layers of glass and magnetite have revealed an important transport effect. The characterization of altered powders revealed a heterogeneous alteration of glass powder: glass grains directly in contact with magnetite powder are more altered than the other grains (figure 9).

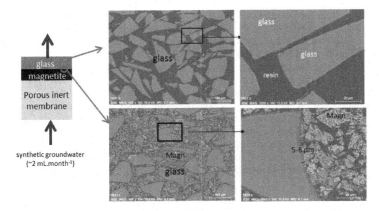

Figure 9. Example of experiment performed in a reactor cell through a successive layer of magnetite and glass powders.

Glass/iron/claystone interactions

Similar diffusion experiments have been performed through successive layers of claystone, iron powder or a punched iron disk, and glass powder [25] in order to study interactions between silicon released by glass alteration and iron released by corrosion. Even though the solution analyses didn't reveal any significant differences between the different configurations, the characterization of altered powders revealed a heterogeneous alteration of glass powder as previously observed in the experiment with glass and magnetite powders. The characterization also allowed to determine the nature of the neoformed phases that are mainly iron silicates and magnesium silicates. Most of the iron silicates were found close to the iron source (powder or disk) and they affected the glass alteration only locally. Glass grains farther from the iron source were less altered. We can then conclude that the precipitation of iron silicates depends on the iron concentration and/or the rate at which it is supplied by corrosion.

The characterization of altered glass also revealed a penetration of iron into the gel. De Combarieu et al. [26] already suggested that the gel formed at the surface of glass was affected by iron that may alter its protective properties. Several recent detailed characterizations of alteration products with Scanning Transmission X-ray Microscopy (STXM) allowed to specify the structural environment of iron on nanometric scale, and to determine the iron speciation both in corrosion products and in alteration layer. Such characterization has been performed both on vitrified by-products from blast furnace (archaeological slag analogs buried 450 years in anoxic iron-rich carbonated conditions) [27-28], and on SON68 glass altered 2 years at 50 °C in presence of iron and claystone [25]. It appears that iron penetrates the alteration layer. According to STXM analyses, its redox state consists of a mix of Fe(II) and Fe(III), and the Fe(II)/Fe(III) ratio is comparable to that in the iron silicates formed close to the alteration layer [25,27-28].

Integrated in situ experiments are also being performed in the Andra's underground research laboratory (Bure, France) in order to study the long-term performance of SON68 glass in realistic disposal conditions. One of these experiments aims at improving knowledge on glass/iron/claystone interface reactivity over multi-annual periods in different configurations. The experimental setup is made of five identical vertical descending boreholes filled with claystone pieces where four different samples are inserted: e.g. a mixture of glass and iron powder, a mixture of glass/iron/claystone powders, glass powder in a carbon steel container, fractured glass block in a carbon steel container (Figure 10)... The five test intervals that were installed in December 2010 have been progressively saturated with the porewater of the surrounding Callovo-Oxfordian argillaceous rock. The first borehole was dismantled in December 2012. After less than 2 years of alteration, the mean glass alteration rate remains high, close to the initial rate, in several configurations revealing that mixtures of powders (glass/iron or glass/iron/claystone) are highly reactive. The next borehole dismantling will allow a better evaluation of the long-term evolution of alteration and to get a better understanding of the importance of reactive surfaces and more specifically of the glass surface over the iron surface ratio on glass alteration.

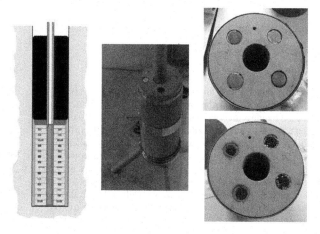

Figure 10. Experimental set up of an in situ experiment performed in the Andra's URL and examples of samples placed in claystone pieces.

OUTLOOK

The studies carried out since 2006 have underlined the important effect of environmental conditions (vapor hydration, clayey water, iron, corrosion products and claystone) on glass alteration. Consequently, our understanding of glass alteration under such conditions has been improved even though some issues remain to be addressed.

Concerning glass vapor hydration, the main problems yet to be addressed are the long-term hydration rate and the radiolysis effect on glass hydration in anoxic conditions. As regards to glass alteration in saturated conditions, in presence of iron or corrosion products, it is still necessary to identify the nature of iron silicate phases (Si/Fe ratio) and to determine the sorption capacity of *in situ* formed corrosion products. Improving reactive transport models is another challenge we will have to take up in the years to come. Indeed improved reactive transport models are required to compute pH and magnesium availability in repository conditions as well as to assess the influence of geometry - glass over iron surface ratio - on glass alteration in order to upscale our experimental results at the waste packages scale.

ACKNOWLEDGEMENTS

This work is the result of a partnership between Andra, CEA, SUBATECH, EDF, LAPA, and LEM in the framework of the Glass/Iron/Clay Laboratory Group program.

REFERENCES

1. Andra , Dossier 2005. Evaluation of the feasibility of a geological repository in an argillaceous formation, http://www.andra.fr/international/download/andra-international-en/document/editions/266va.pdf (2005)
2. J. Neeway, *PhD Thesis*, University of Nantes (2011)
3. J. Neeway, A. Abdelouas, B. Grambow, S. Schumacher, C. Martin, M. Kogawa, S. Utsunomiya, S. Gin, P. Frugier, *J. Non-Cryst. Solids* 358, 2894-2905 (2012)
4. R. Bouakkaz, *PhD Thesis*, University Nantes Angers Le Mans, France (2014)
5. W.L. Gong, L.M. Wang, R.C. Ewing, E. Vernaz, J.K. Bates, W.L. Ebert, *J. Nucl. Mater.* 254, 249-265 (1998)
6. A. Ait Chaou, A. Abdelouas, Y. El Mendili, R. Bouakkaz, C. Martin, *Procedia Materials Science* (in press)
7. P. Jollivet, S. Gin, S. Schumacher, *Chem. Geol.* 330-331, 207-217 (2012)
8. P. Jollivet, P. Frugier, G. Parisot, J.P. Mestre, E. Brackx, S. Gin, S. Schumacher, *J. Nucl. Mater.*, 420, 508-518 (2012)
9. P. Frugier, S. Gin, Y. Minet, T. Chave, B. Bonin, N. Godon, J.-E. Lartigue, P. Jollivet, A. Ayral, L. De Windt, G. Santarini, *J. Nucl. Mater*, 380, 8-21 (2008)
10. K. Lemmens, *J. Nucl. Mater.*, 298, 11-18(2001).
11. M.L. Schlegel, C. Bataillon, K. Benhamida, C. Blanc, D. Menut, and J.-L. Lacour , *Appl. Geochem.* 23, 2619-2633 (2008)
12. M.L. Schlegel, C. Bataillon, C. Blanc, D. Prêt, E. Foy, *Environ. Sci. Technol.*, 44, 1503-1508 (2010)
13. S. Necib, D. Crusset, N. Michau, F. Foct, M.L. Schlgel, S. Daumas, S. Dewonck, *Eurcorr European Corrosion Congress* (2014)
14. D. Neff, M. Saheb, J. Monnier, S. Perrin, M. Descostes, V. L'hostis, D. Crusset, A. Millard, P. Dillmann, *J. Nucl. Mater.*, 402, 196-205 (2010)
15. M. Saheb, D. Neff, P. Dillmann, H. Matthiesen, and E. Foy, *J. Nucl. Mater.*, 379, 118-123 (2008)
16. M. Saheb, D. Neff, J. Demory, E. Foy and P. Dillmann, *Corros. Eng. Sci. Technol.* 45, 381-387 (2010a).
17. M. Saheb, M. Descostes, D. Neff, H. Matthiesen, A. Michelin and P. Dillmann, *Appl. Geochem.* 25, 1937-1948 (2010b).

18. B. Grambow, H.U. Zwicky, G. Bart, I.K. Björner and L.O. Werme, *Mat. Res. Soc. Symp. Proc.*, 84, 471-481 (1987).
19. G. Bart, H.U. Zwicky, E.T. Aerne, Th. Graber, D. Z'Berg and M. Tokiwai, *Mat. Res. Soc. Symp. Proc.*, 84, 459-470 (1987)
20. V. Philippini, A. Naveau, H. Catalette and S. Leclercq, *J. Nucl. Mater.*, 348, 60-69 (2006).
21. N. Jordan, N. Marmier, C. Lomenech, E. Giffaut and J. Ehrhart, *J. Colloid Interface Sci.*, 312,(2), 224-229 (2007).
22. C. Mayant, B. Grambow, A. Abdelouas, S. Ribet and S. Leclercq, *Phys. Chem. Earth*, 33, 991-999 (2008).
23. N. Godon, S. Gin, D. Rebiscoul, and P. Frugier, *Procedia Earth and Planetary Science*, 7, 300-303 (2013)
24. J. van der Lee, L. De Windt, V. Lagneau, P. Goblet, *Comput. Geosci.* 29, 265–275 (2003)
25. E. Burger, D. Rebiscoul, F. Bruguier, M. Jublot, J.-E. Lartigue, S. Gin, *Appl.Geochem.*, 31, 159-170 (2013)
26. G. De Combarieu, M.L. Schlegel, D. Neff, E. foy, D. Vantelon, P. Barboux, S. Gin, *Appl. Geochem.*, 26, 65-79 (2011)
27. A. Michelin (2011), *PhD Thesis*, University of Paris VI – Pierre et Marie Curie.
28. A. Michelin, E. Burger, D. Rebiscoul, D. Neff, F. Bruguier, E. Drouet, P. Dillmann, S. Gin, *Environ. Sci. Technol.*, 47, 750-756 (2013)

Mater. Res. Soc. Symp. Proc. Vol. 1744 © 2015 Materials Research Society
DOI: 10.1557/opl.2015.503

Glass Corrosion in the Presence of Iron-Bearing Materials and Potential Corrosion Suppressors

Joelle Reiser[1], Lindsey Neill[1], Jamie Weaver[1], Benjamin Parruzot[1], Christopher Musa[1], James Neeway[2], Joseph Ryan[2], Nikolla Qafoku[2], Stéphane Gin[3], Nathalie A. Wall[1]

[1] Washington State University, Chemistry Department, Pullman, WA 99164, USA
[2] Pacific Northwest National Laboratory, Energy and Environment Directorate, Richland, WA 99352, USA
[3] CEA Marcoule DTCD/SECM, F-30207 Bagnols-sur-Cèze, France

ABSTRACT

A complete understanding of radioactive waste glass interactions with near-field materials is essential for appropriate nuclear waste repository performance assessment. In many geologic repository designs, Fe is present both in the natural environment and in the containers that will hold the waste glasses. In this paper we discuss investigations of the alteration of International Simple Glass (ISG) in the presence of Fe^0 foil and hematite (Fe_2O_3). Based on solid analysis, ISG alteration is more pronounced in the presence of Fe^0 than with hematite. Additionally, typical glass corrosion is observed for distances of 5 mm between Fe materials and ISG, but incorporation of Fe in the alteration layer is only observed for systems exhibiting full contact between Fe^0 material and ISG. Solution analysis results indicate that diatomaceous earth minimizes corrosion to a larger extent than fumed silica does when present with iron and ISG.

INTRODUCTION

The understanding of the long-term evolution of borosilicate glasses is of particular importance for the disposal of high-level radioactive waste (HLW) [1, 2, 4], and glass alteration in the presence of near-field materials must be known [2, 3]. Iron is a particularly important near-field material since it is in the natural environment and in the stainless steel containers that store the waste glass [2]. Iron also is a component that is present in many waste glasses. When in the 2+ oxidation state in solution, iron is known to accelerate the glass alteration rate but the corresponding mechanisms are not fully understood [3]. This work presents experimental results regarding 1) glass alteration in the presence of Fe^0 and Fe_2O_3 to evaluate the effect of the Fe oxidation state on glass alteration and 2) the efficacy of potential glass corrosion suppressants at 90°C: diatomaceous earth (DE) and fumed silica (FS). International Simple Glass (ISG) was used for these studies; ISG, a six-component borosilicate glass, was developed as a reference benchmark glass for an international collaboration on waste glass alteration mechanisms [4].

EXPERIMENT

Materials

International Simple Glass (ISG) has a density of 2.5 $g \cdot cm^{-3}$ and a composition in weight percentages as follows: SiO_2: 56.2%, B_2O_3: 17.3%, Na_2O: 12.2%, Al_2O_3: 6.1%, CaO: 5.0% and ZrO_2: 3.3% [5]. The glass was obtained in bar form from Savannah River National Laboratory. Coupons were cut with a low-speed saw with a diamond tipped blade (Buehler Isomet®). The coupons were polished on a Leco Grinding and Polishing® SS-200 polisher up to 1200 grit. Dimensions of each coupon were approximately $10 \times 20 \times 2$ mm^3. Hematite coupons (Fisher®) were cut to the same dimensions as the glass coupons and polished to 320 to 600 grit. Fe^0 foil (Sigma Aldrich®) of thickness 0.1 mm was cut to two of the same dimensions as the coupons 10×20 mm^2. Diatomaceous earth (DE) (MP-Biomedicals®) and fumed silica (SiO$_2$, called FS hereafter) (Sigma Aldrich®) were prepared as powders; DE was sieved to $149 - 250$ µm, and FS was available at $0.2 - 0.3$ µm. Deionized distilled water (18 MΩ) (DIW) was used.

Assembly Description

Figure 1 presents the experimental setups. The configuration in Figure 1a allows for determination of ISG corrosion in the presence of iron sources and as a function of the distance between the iron source and glass (both 0 mm – i.e. full contact – and 5 mm). Previous studies showed the importance of a complete contact between the iron source and the glass coupons, which allows for uniform glass corrosion [3]. In the present work, Fe^0 foil was etched with a knife in a grid pattern to facilitate solution flow between ISG and the Fe^0 foil. The etched Fe^0 foil was flattened with a metal roller onto the ISG coupon, and the polished hematite coupon was pressed against the ISG surface. The gap between the two ISG coupons was maintained by 5 mm Teflon® spacers, and the sample configurations were held together with chemically inert glue (Loctite® Plastics Bonding System). Solution was allowed to flow along every surface of the glass except the portions touching the Teflon® blocks. Additional Teflon® spacers were used to elevate the bottom ISG coupon from the experimental stainless steel vessel (50 mL Parr®). The efficacy of glass corrosion suppressant (CS) – DE or FS – in the presence of Fe^0 was tested using the set-up presented in Figure 1b. The CS was placed in a bag made of 30 micron nylon mesh (Industrial Netting®) between the Fe^0 foil and the top ISG coupon, while the bottom ISG coupon remained out of contact with the CS. Solution was allowed to flow through the mesh bag to allow for every glass surface to be in contact with solution with the exception of portions in contact with the Teflon® blocks. Additionally, blank samples were prepared: Teflon® blocks and glue; ISG with Teflon® blocks and glue; FS with Teflon® blocks and glue, DE with Teflon® blocks and glue; ISG with DE, Teflon® blocks, and glue; and ISG with FS, Teflon® blocks, and glue. The following nomenclature is used to designate experimental setups: ISG/iron source/CS (e.g. ISG/Fe/DE means ISG alteration in the presence of Fe^0 and DE, ø indicates the absence of iron source or CS). Each sample set was prepared in duplicates for statistical analysis.

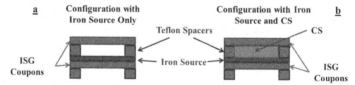

Figure 1: Schematic Representation of the Experimental Configurations with a) Iron Source Only and b) Iron Source and Corrosion Suppressant (CS)

Each configuration was inserted into Parr® vessels containing 25 mL DIW to obtain a glass-surface-area-to-solution-volume ratio (S/V) of 40 m^{-1} with the exception of ISG/ø/ø which had an S/V of 20 m^{-1}. The solutions were purged with pure N$_2$ (A-L Compressed Gases, Inc.) and solution pH was measured at room temperature before sealing the Parr® vessels. All experiments were carried out in a 90°C oven for 1 month without disturbing the samples. Losses due to evaporation were negligible. The solutions were then removed and pH was measured at room temperature. Glass samples were washed with ethanol and dried in a desiccator. ISG samples were then cut for various analyses.

Characterization Techniques

Solutions were diluted with 1-2% HNO_3 and analyzed using Inductively Coupled Plasma Optical Emission Spectrometry (ICP-OES) (Perkin Elmer® Optima 3200 RL) for Al, B, Ca, Fe, Na, and Si; the instrument was calibrated using dilutions of standard solutions (Inorganic Ventures).

Normalized loss (NL_x) of B, Si, and Na from ISG in $g·m^{-2}$ was calculated using the following equation:

$$NL_x = \frac{C_x}{(\frac{S}{V})f_x} \qquad (1)$$

Where C_x is the elemental concentration in $g·m^{-3}$, f_x is the mass fraction of the selected element in ISG, and S/V in m^{-1}. NL(Si) is corrected for excess Si for samples that contained fumed silica, based on Si concentration determined in ø/ø/FS configurations. This correction is calculated from the Si release of a sample of fumed silica altered in water for one month. This correction could not be applied to the samples containing DE due to experimental error associated with the Si concentration for the ø//ø/DE blank.

Altered monoliths were embedded in epoxy resin (Specifix-20®) polished to 1200 grit and sectioned to expose the cross section. Altered and unaltered glass compositions were measured by Energy-Dispersive X-ray Spectrometry (EDS) using the JEOL JXA-8500F electron microprobe, equipped with a Thermo Scientific UltraDry EDS detector and ThermoNORAN™ System 7 analytical software. Measurements were made using an accelerating voltage of 15 kV, and a beam current of 8 nA. The beam was defocused to a 1μm spot size to help mitigate alkali migration under beam irradiation [6]. Due to the roughness of the surfaces, quantitative data could not yet be obtained from EDS analyses. The thickness of the alteration layer was measured using the SEM images of the cross sections and ImageJ software. Alteration layer thickness was measured at multiple points on a glass side to allow for statistical analysis.

DISCUSSION

Over the course of the experiment, sample solution pH increased from a range of 5.0 - 6.0 to 7.5 - 9.5. Figure 2 shows NL_x for B, Na, and Si. Uncertainty is reported as twice the standard deviation associated with results obtained from duplicate analyses. Boron is used as a tracer of glass alteration because B is known to not be retained in the alteration products. NL(B) and NL(Na) are comparable for most samples for the chosen S/V of 40 m^{-1}. NL(Si) is smaller than NL(B) and NL(Na) for each sample; NL(Si) is particularly small for ISG/Fe/ø. In general, samples featuring corrosion suppressants, FS or DE, have lower NL(B) than their counterparts in absence of FS or DE, indicating that corrosion suppression is occuring. However, there is no statistical difference between NL(B) values for ISG/Fe/ø and ISG/Fe/FS. At this point, FS is not known to be an appropiate suppressant for ISG corrosion in the presence of Fe^0. NL(B) is slightly smaller for ISG/Fe/DE than for ISG/Fe/FS.

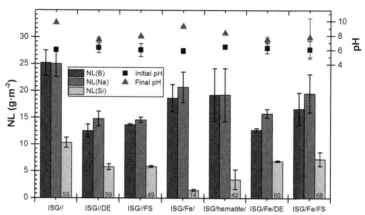

Figure 2: Normalized mass losses (NL_x) for each sample calculated from solution concentrations of B, Na, and Si (left axis), initial and final solution pH (right axis), and corrected concentrations of Si in ppm shown in the NL(Si) bars

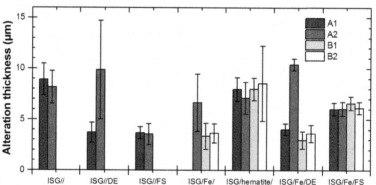

Figure 3: Alteration thickness measurements obtained from SEM cross sections of altered coupons. The labels indicate the identity of the glass coupon and its side and refer to the drawing, with A1 being the top side of the upper glass coupon, for example. SEM micrographs will be provided in another manuscript.

The alteration layer thicknesses of the various surfaces are shown in Figure 3; alteration layers were not observed on all surfaces because the layers were too thin. Additionally, the surface in contact with iron (B1) in ISG/Fe/DE, ISG/Fe/FS, and ISG/Fe/ø configurations had alteration layers that contained iron as indicated by EDS spectra. A2 alteration layers did not contain iron, but secondary products featuring trace levels of iron were observed at the surface of the altered glass coupon. Within a single sample, alteration layer thicknesses are not identical within reported uncertainty for ISG/ø/DE (A1 and A2) and ISG/Fe/DE (A2 versus all the other coupon sides). For all the other samples, alteration thicknesses do not vary between glass coupon sides. However, for the surface of the top coupon in contact with the mesh bag that contains the CS, the alteration thickness is larger for ISG/Fe/FS than for ISG/Fe/DE for surfaces B1 and B2, and the alteration thickness is larger for ISG/hematite/ø than for ISG/Fe/ø. Comparison between ISG/ø/ø and ISG/Fe/ø shows no statistical difference between their results at A2, indicating that the upper coupon is too far from the Fe source for Fe to influence alteration. .

NL$_x$ values determined from solution analysis can be converted to an equivalent altered glass thickness based on the density of ISG. This equivalent altered glass thickness is an average value over the whole glass surface area, S, and can thus be compared to the average of the 4 alteration layer thicknesses obtained from solid state analysis measurements (A1, A2, B1 and B2). Figure 4 shows the comparison of alteration layer thickness from solution and solid state analysis. The dashed columns on Figure 4 represent configurations that featured Fe in the some alteration layers. For configurations ISG/ø/ø, ISG/ø/DE, ISG/Fe/DE, and ISG/Fe/FS, the calculated solution and measured solid alteration thicknesses are statistically equivalent.

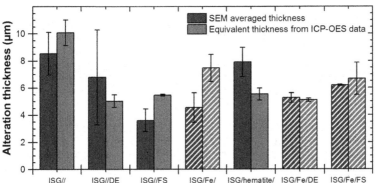

Figure 4: Alteration Thickness Comparison. Dashed columns: presence of Fe in some alteration layers.

The alteration thickness for ISG/Fe/ø measured by SEM is statistically smaller than the thickness determined from solution data; this may be due to calculation artifacts that do not account for the fact that the presence of Fe in the alteration layer increases the layer density. The differences between the resulting alteration thicknesses for ISG/hematite/ø can be due to an excessive amount of glue on the surface of the coupons that reduced the glass surface area in contact with water. The apparent differences observed for the ISG/ø/FS alteration thicknesses deducted from either solution or solid analysis will be investigated further with additional long-term experiments.

CONCLUSIONS

Results presented in this paper are preliminary data for further work, in which similar glass corrosion tests will be run for longer time periods and the distance between upper glass coupons and Fe sources will be decreased. Additionally, the phases into which Fe is incorporated in the alteration layer will be investigated.

ACKNOWLEDGMENTS

This research is being performed using funding received from the DOE Office of Nuclear Energy's Nuclear Energy University Programs, under Project 23-3361.The authors would like to thank Dr. Owen Neill of the WSU School of the Environment for his assistance with the electron microprobe solid state analysis.

REFERENCES

[1]   E. Burger, D. Rebiscoul, F. Bruguier, M. Jublot, J. E. Lartigue, and S. Gin, "Impact of iron on nuclear glass alteration in geological repository conditions: A multiscale approach," *Appl. Geochemistry*, vol. 31, pp. 159–170, Apr. 2013.

[2]   B. Fleury, N. Godon, A. Ayral, and S. Gin, "SON68 glass dissolution driven by magnesium silicate precipitation," *J. Nucl. Mater.*, vol. 442, no. 1–3, pp. 17–28, Nov. 2013.

[3]   A. Michelin, E. Burger, E. Leroy, E. Foy, D. Neff, K. Benzerara, P. Dillmann, and S. Gin, "Effect of iron metal and siderite on the durability of simulated archeological glassy material," *Corros. Sci.*, vol. 76, pp. 403–414, Nov. 2013.

[4]   S. Gin, a. Abdelouas, L. J. Criscenti, W. L. Ebert, K. Ferrand, T. Geisler, M. T. Harrison, Y. Inagaki, S. Mitsui, K. T. Mueller, J. C. Marra, C. G. Pantano, E. M. Pierce, J. V. Ryan, J. M. Schofield, C. I. Steefel, and J. D. Vienna, "An international initiative on long-term behavior of high-level nuclear waste glass," *Mater. Today*, vol. 16, no. 6, pp. 243–248, Jun. 2013.

[5]   Y. Inagaki, T. Kikunaga, K. Idemitsu, and T. Arima, "Initial Dissolution Rate of the International Simple Glass as a Function of pH and Temperature Measured Using Microchannel Flow-Through Test Method," *Int. J. Appl. Glas. Sci.*, vol. 4, no. 4, pp. 317–327, Dec. 2013.

[6]   C. H. Nielson and H. Sigurdsson, "Quantitative methods for electron microprobe analysis of sodium in natural and synthetic glasses," *Am. Mineral.*, vol. 66, pp. 547–552, 1981.

Mater. Res. Soc. Symp. Proc. Vol. 1744 © 2015 Materials Research Society
DOI: 10.1557/opl.2015.331

Uncertainty in the Surface Area of Crushed Glass in Rate Calculations

William L. Ebert[1], Charles L. Crawford[2], and Carol M. Jantzen[2]
[1]Argonne National Laboratory,
Argonne, IL, U.S.A
[2]Savannah River National Laboratory,
Aiken, SC, U.S.A.

ABSTRACT

Series of 7-day Product Consistency Tests (PCTs) were conducted with ARM-1 glass using the -100+200 mesh size fraction and several sub-fractions to measure the sensitivity of the test response to the distribution of particle sizes. Separate samples were prepared for testing by dry sieving and wet sieving, and the particle size distributions and PCT responses were measured for each fraction. Triplicate tests were conducted at 90 °C using a water/glass mass ratio of 10.0 with each size fraction. Test results are evaluated regarding the sensitivity of the test response to the particle size distributions and, conversely, the uncertainty due to calculating the surface areas (and dissolution rates) by modeling the particles as spheres. These analyses show the solution feedback effects of dissolved glass constituents (i.e., the reaction affinity) counteract the effects of the glass surface areas provided by different particle size distributions on the test response. The opposing effects of the surface area on the amount of glass dissolved and on the glass dissolution rate moderate the sensitivity of the PCT response to the particle size distribution.

INTRODUCTION

Experiments were conducted to address the uncertainty in dissolution rates calculated from tests using crushed glass following the general Product Consistency Test (PCT) procedure [1]. The glass dissolution rates determined using several test methods with crushed glass are usually represented on a per-area basis to compare the extents of dissolution of different glasses under various test conditions and to upscale the measured rates to represent full-sized glass waste forms in performance calculations. Samples of crushed glass are prepared for PCTs by isolating the desired size fraction by sieving, and the specific surface area is estimated by relating the mean sieve opening to the dimension of a geometric model of the particle, such as the diameter of a sphere or edge of a cube. The tests and analyses presented herein address the effect of the distribution of particle sizes in a test sample on the calculated dissolution rate relative to the uncertainty in the test response and repeatability of the test. Previous analyses have shown the effective particle dimensions of crushed glasses in the -100 +200 mesh size fraction (msf)[1] prepared for tests such as the PCT have a nearly normal (Gaussian) size distribution [1]. This means the surface area available in the PCTs, which is proportional to the square of the particle dimension, will be skewed towards the contributions of smaller particles in a test sample based on mass. Variances in the particle distributions that originate during the crushing, sieving, and washing steps to prepare the glass will affect the surface area available in a PCT. The dissolution rate is calculated from the amount of glass that dissolves, based on the solution concentrations of soluble glass constituents, normalized by the glass surface area and test duration. The single value of the specific surface area that is selected to represent the entire size fraction should

---

[1] This is the fraction of particles that pass through a 100 mesh sieve but not through a 200 mesh sieve.

minimize the error in the calculated the dissolution rate due to the irregular particle shapes and the particle size distribution. Although the average geometric dimension based on the sieve fraction may not provide the best representation of the surface area, the error it imposes should be within the uncertainty of the test method. These tests were conducted using samples having particle size distributions that were intentionally-biased to evaluate the sensitivity of the PCT response and the approach for estimating the surface area to calculate the glass dissolution rate.

EXPERIMENT

ARM-1 glass was crushed and dry-sieved at ANL using a mechanical sifter to isolate the -100 +200 msf, which was then repeatedly washed to remove fines by agitation and decanting. A portion of that glass was sieved again (dry) to isolate the -100 +120, -120 +140, and -140+200 msf, which were again washed to remove fines and dried. A separate batch of the same ARM-1 glass was prepared at SRNL by flushing crushed glass through a series of sieves with demineralized water to isolate the -100 +200 msf. A portion of that glass was then re-sieved (wet) to isolate the -100 +120, -120 +140, and -140 +200 msf. Samples of the dry-sieved and wet-sieved materials were analyzed with SEM and a Microtrac particle size analyzer to measure the particle size distributions in each size fraction. Figure 1 presents SEM photomicrographs of the dry-sieved size fractions showing the general size, shape, and surface conditions of the different fractions; a 200 x 200 μm square is drawn around a particle in each figure to facilitate

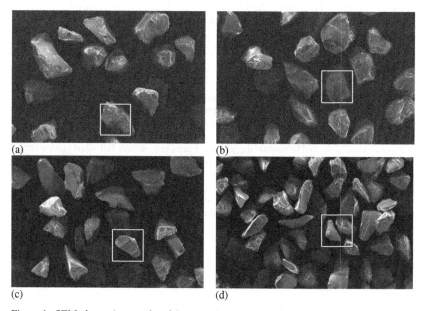

(a)   (b)

(c)   (d)

Figure 1. SEM photomicrographs of dry-sieved ARM-1 glass in (a) -100+200, (b) -100+120, (c) -120+140, and (d) -140+200 mesh size fractions. The boxes represent 200 x 200 μm.

comparisons. Both the dry-sieved and wet-sieved materials were meticulously washed to remove fines. The appearance and sizes of the wet-sieved and dry-sieved particles were indistinguishable by SEM examination.

In the PCT method, a known mass of crushed glass within a particular size fraction is immersed in an amount of water proportional to the mass of glass at an elevated temperature for predetermined duration [1]. The concentrations of soluble glass constituents measured in the test solution are used as a measure of the mass of glass that dissolved, based on the concentrations of those constituents in the glass and the solution volume. The mass of glass that dissolves can be further normalized to the glass surface area and represented as the normalized mass loss, which is calculated as

$$NL(i) = \frac{C(i) \times (m_w / \rho_w)}{S_{sp} \times m_g \times f_i} = \frac{C(i)}{(S/V) \times f_i} = \frac{m(i)}{S \times f_i},$$ (1)

where $C(i)$ is the concentration of species i in the test solution, m_w and m_g are the masses of water and glass used in the test, ρ_w is the density of water, $f(i)$ is the mass fraction of species i in the glass, and S_{sp} is the specific surface area of the crushed glass, S is the glass surface area, V is the solution volume, and $m(i)$ is the mass of species i released. Equation 1 expresses NL(i) in terms of measured [m_w, m_g, f_i, $C(i)$], estimated [S_{sp}], and calculated [S, V, $m(i)$] values. The elemental mass fractions are determined from the composition of the glass, which is given in Table 1. The NL(i) values can be divided by the test duration to calculate an average dissolution rate.

Series of triplicate PCTs were conducted at 90 °C in demineralized water with the different size fractions prepared by dry or wet sieving. The results of tests with dry-sieved glass conducted at ANL and with wet-sieved glass conducted at SRNL were compared separately. All tests were conducted in stainless steel vessels with a water-to-glass mass ratio (m_w/m_g) of 10.0 for all size fractions to directly compare the test responses and evaluate the effect of S_{sp} values calculated for the different size distributions. The mass ratio rather than the S/V ratio was kept constant to evaluate the effect of the particle size distribution (which is assumed to be proportional to the total surface area) on the test response. Direct comparisons indicate the sensitivity of the PCT response to unintentional upsets that could occur during sample preparation that impact the glass particle size distribution and the surface area that is available in the test. The responses of tests with different size fractions were compared with the experimental precision determined in tests with the -100 +200 msf.

Table 1. Composition of ARM-1 glass, oxide mass%

Oxide	ARM-1	Oxide	ARM-1	Oxide	ARM-1
Al_2O_3	5.59	Li_2O	5.08	SrO	0.45
B_2O_3	11.30	MoO_3	1.66	TiO_2	3.21
BaO	0.66	Na_2O	9.66	ZnO	1.46
CaO	2.24	Nd_2O_3	5.96	ZrO_2	1.80
CeO_2	1.51	P_2O_5	0.65		
Cs_2O	1.17	SiO_2	46.50	Sum	98.90

RESULTS

The Microtrac software provides distributions based on diameter, area, and volume, but only the measured distributions of particle diameters are used in this evaluation. The Microtrac results for the distribution of particle diameters are plotted in Figure 2a for all the prepared glasses. (The distributions of all size fractions are normalized to 100% occurrence). The vertical dashed lines indicate the opening sizes of the sieve screens. Although the particle size distributions of the dry-sieved and wet-sieved materials are essentially identical for the intermediate size fractions, the wet-sieved -100 +200 msf has a greater abundance of smaller particles than the dry-sieved distribution. The ranges of particle diameters exceed the ranges of sieve openings used to separate each fraction. Figure 2b shows the Microtrac results for the distribution of particle diameters for the dry-sieved -100 +200 msf and Gaussian fit representing the particle diameters and the distribution of particle surface areas modeled as spheres having those diameters. The distribution of particle surface areas is proportional to the square of the diameter and further skewed to smaller areas. The double-headed horizontal arrow in Figure 2b shows the range of sieve openings and the vertical arrows (left to right) identify the average of the openings and the mean area and diameters of the particle distributions. Although the peak of the diameter distribution (at about 132 μm) occurs at a significantly higher value than the average of the -100 +200 mesh opening, which is about 112 μm, the diameter corresponding to the peak of the area distribution (at about 119 μm) is well-represented by that average value. That is, the average mesh opening does not represent the Gaussian mean of the distribution of particle diameters very well, but it does represent the Gaussian mean of the distribution of the calculated surface areas well. The distributions of particle diameters for all size fractions of the dry-sieved materials had a Gaussian size distribution that is slightly skewed towards smaller particles. That is, there are more large particles than predicted by the Gaussian distribution for each fraction.

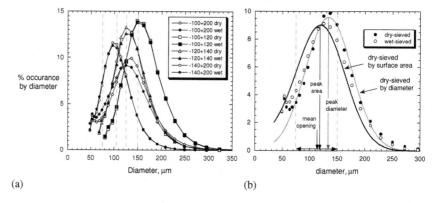

(a)   (b)

Figure 2. (a) Microtrac results for different size fractions of dry- and wet-sieved ARM-1 glass and (b) Gaussian fits of Microtrac results for -100+200 mesh size fractions of dry-sieved ARM-1 glass and calculated distribution of areas for spherical particles.

148

The specific surface area for each size fraction was calculated by modeling the particles as spheres using the equation

$$S_{sp} = 6/(d \bullet \rho_g),$$ (2)

where d is the particle diameter set equal to the arithmetic average of the sieve openings for the size fraction and ρ_g is the density of ARM-1 glass, which is 2.75 g cm^{-3}. The values are summarized in Table 2, including the glass surface area-to-water volume (S/V) ratios for each size fraction for the targeted water-to-glass mass ratio of 10.0 and the ratios of the specific surface areas and S/V ratios of the sub-fractions to those values for the -100 +200 msf. The S/V ratios are 19% lower for the -100 +120 msf and 25% higher for the -140+200 msf than the S/V ratio for the -100+200 msf based on spherical particles. The S/V ratios are about the same for the -120+140 and -100+200 msf.

Normalized mass loss values were calculated from the measured solution concentrations in two ways: first using the specific surface area of the -100 +200 msf to evaluate the sensitivity of the PCT to the imposed variations in the particle size distributions and then using the specific surface areas calculated for each sub-fraction to evaluate that method. The spherical model was used in all cases. The NL(i) values for tests with the sub-fractions are compared with the values for tests with the -100 +200 msf in both methods. The tests with dry-sieved glass conducted at ANL and with the wet-sieved glass conducted at SRNL are evaluated separately based on the within-laboratory test precision determined for each laboratory. The repeatability of tests conducted at each laboratory with the -100 +200 msf glass is expected to be within the respective ranges 99.7% of the time based on normal distributions of all factors contributing to the test variance. This defines the repeatability range, and values for tests conducted with the sub-fractions that fall outside this range are statistically different and distinguished by the PCT.

The mean values plus and minus three standard deviations (3s) for tests with the dry- and wet-sieved -100 +200 msf are shown as horizontal dashed lines in Figure 3. Figure 3a shows NL(i) calculated using the same S_{sp} for all fractions (0.0207 m^2 g^{-1}) and Figure 3b shows NL(i) calculated with S_{sp} representing the individual sub-fractions (from Table 2). The heights of the bars give the mean values and the uncertainty bars represent ± 3s. Small "+" or "-" symbols included on the bars in Figures 4a and 4b indicate ranges of responses (mean ± 3s) for a particular element and sub-fraction that fall above or below the repeatability range.

Table 2. Specific surface area of spherical particles based on mesh size fraction

Mesh size	Mesh opening, mm	Mesh size fraction	Average dimension, m	Specific Surface Area, m^2/g	Target S/V, m^{-1}	Ratio[a]
100	0.149	-100+200	1.12E-04	0.0207	2070	1.00
120	0.125	-100+120	1.37E-04	0.0168	1684	0.81
140	0.105	-120+140	1.15E-04	0.0201	2007	0.97
200	0.074	-140+200	8.95E-05	0.0258	2578	1.25

[a] Ratio of specific surface area and S/V ratio to that of -100+200 size fraction.

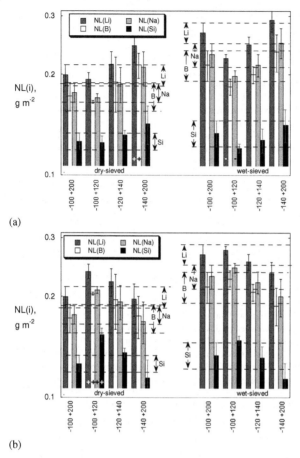

(a)

(b)

Figure 3. Results of triplicate PCT calculated (a) using S_{sp} of the -100 +200 mesh size fraction for all size fractions and (b) using the S_{sp} of each sub-fraction.

In Figure 3a, only two responses in the tests with the smallest size fraction of dry-sieved glass and two responses in the tests with the largest size fraction of wet-sieved glasses lie outside the repeatability ranges. These results show the PCT-A response is not sensitive to differences in the particle size distributions of the different mesh size fractions. Figure 3b shows all responses in tests with the -100 +120 msf of dry-sieved glass lie outside the repeatability range, but the responses of other dry-sieved fractions and all test responses with wet-sieved glass lie within the repeatability range for tests conducted within the same laboratory.

Notice that the mean values of all NL(i) increase with decreasing sub-fraction size in Figure 3a but decrease in Figure 3b. This is attributed to competing effects of surface area and reactivity in PCTs with the different size fractions. NL(Si) has the greatest positive deviations for the largest size fractions and greatest negative deviations for the smallest size fractions for both the dry- and wet-sieved glass. The smallest deviation in NL(Si) occurs in tests with the -120 +140 msf. The release of Si indicates the relative effects of solution feedback in the tests with different size fractions, which is proportional to $C(Si)$. Comparing the relative differences in S/V ratios for the different sub-fractions in Table 2 with the differences in NL(Si) shows that much of the difference in the tests responses can be attributed to the different $C(Si)$ that are attained in the tests. Since the masses of water and glass were essentially the same in all tests and the same glass was used, differences in NL(i) values calculated with Equation 1 are due to the values of the specific surface area used to calculate S for each size fraction and the value of $C(Si)$ attained in each test. For example, the decreased amount of glass available to dissolve due to the 19% lower surface area available in tests with the -100 +120 msf of glass compared to tests with the -100 +200 msf is partially compensated for by the lower attenuations of the rate due to solution feedback in those tests. Likewise, the greater amount of glass available to dissolve due to the 25% higher surface area available with the -140 +200 msf of glass is partially compensated for by the greater attenuations due to solution feedback in tests with the smaller particles.

DISCUSSION

Sized test materials prepared by dry or wet sieving have equivalent size distributions. This is based on the similarities of the sub-fractions prepared from the -100+200 msf material. The minor difference seen in the dry- and wet-sieved -100+200 msf is considered off-normal based on the observed deviations from a Gaussian distribution. It is likely that bias in the crushed source material that is sieved will bias the size distributions in the sieved fractions. For example, repeated crushing-sieving cycles performed with the same batch of glass during sample preparation are expected to bias the fractions that are separated toward smaller particles. This could be an issue when only small amount of glass is available for testing.

The effective ranges of particle sizes in the different fractions that result from using different sieve sizes exceed the range of sieve openings. Further constraining the sieve sizes will not produce a significantly smaller range of particle sizes. Based on the spherical particle model, the overlaps with the -100 +200 msf were 48% and 46% for the -100 +120 and -140 +200 msf in terms of surface area. These sub-fractions are otherwise biased to lower and higher surface areas, respectively. The overlap of the -100 +120 and -140 +200 msf was 28%. The -120 +140 msf has about the same mean as the -100 +200 msf but a narrower distribution.

The responses of PCTs are mostly insensitive to variances in the size fraction that may occur during sample preparation for tests conducted within the range of acceptable test parameters and the repeatability is about the same. The relationships between the PCT responses, particle size distributions, sample preparation methods (dry or wet sieving), and water-to-glass mass ratio are due primarily to the S/V ratio imposed by the test conditions. These tests show the effect of increased glass surface area to increase the amount of glass dissolved is counteracted by the effect of the reaction affinity resulting from the increasing concentration of (primarily) dissolved silica to decrease the dissolution rate. The particle size distribution affects both S and $C(Si)$ that have opposing effects on NL(i) and the glass dissolution rate. Those effects also decrease the sensitivity of the PCT response to the particle size distribution and increase the test precision.

The sensitivity of the PCT response to the test conditions indicates the inherent uncertainty in the relationship between the dissolution rate and the surface area (and the S/V ratio). The results in Figure 3a indicate that the PCT response does not distinguish a difference of 10% in the glass surface area in the test (based on the spherical particle model), which propagates to 10% uncertainty in the average dissolution rate. The sensitivity of the response to the glass surface area is within the uncertainty for to the ranges of testing parameter values used in this study, which were smaller than the allowed ranges for PCT-A. Further analyses of these results and the results of other tests being conducted to evaluate the effects of the water-to-glass mass ratio and test duration over the allowed ranges on the dissolution rates measured with PCTs are in progress. These results will also be evaluated based on the results of an interlaboratory study to measure the reproducibility of PCTs conducted at several laboratories.

CONCLUSIONS

The sensitivity of the PCT response to the surface area is moderated by the coupled and competing effect of the reaction affinity on the dissolution rate, which is more pronounced in tests at high S/V ratios. These test results indicate that the effect of the reaction affinity dominates the effect of the surface area under typical PCT conditions. The effects of varying particle size distribution due to sample preparation are within the testing uncertainty for the ranges of test conditions (e.g., the water-to-glass mass ratio) allowed within the test specifications.

The rate measured in a PCT can be normalized to the surface area estimated from the size fraction used in the test, but this does not address the impact on $C(Si)$ or its effect on the rate. The rates measured with PCT will only be representative of the solution compositions (primarily the silica concentrations) that are attained in those tests and the corresponding reaction affinity.

ACKNOWLEDGMENTS

This work was performed under the auspices of the Materials Recovery and Waste Form Development campaign of the DOE Fuel Cycle Research and Development program. The SEM photomicrographs were provided by J.A. Fortner (ANL). Work at ANL is supported by the U.S. Department of Energy, Office of Nuclear Engineering, Science and Technology under contract W-31-109-Eng-38. Work at SRNL is supported by the U.S. Department of Energy, Office of Environmental Management, EM-31, under Contract DE-AC09-08SR22470 and the U.S. Department of Energy Office of Nuclear Energy, under Contract DE-AC02-06CH11357.

REFERENCES

1. Standard Test Methods for Determining Chemical Durability of Nuclear Waste Glasses: The Product Consistency Test (PCT) Standard C1285-14, Annual Book of ASTM Standards Vol. 12.01, American Society for Testing and Materials, West Conshohocken, PA (2014).

Mater. Res. Soc. Symp. Proc. Vol. 1744 © 2015 Materials Research Society
DOI: 10.1557/opl.2015.332

About U(t) form of pH-dependence of glass corrosion rates at zero surface to volume ratio

Michael I. Ojovan[1] and William E. Lee[2]
[1]Department of Nuclear Energy, IAEA,
Vienna 1020, Austria.
[2] Centre for Nuclear Engineering, Imperial College London,
London SW7 2AZ, UK

ABSTRACT

The pH-dependence of glass corrosion rates has a well-known U-shaped form with minima for near-neutral solutions. This paper analyses the change of U-shaped form with time and reveals that the pH dependence evolves even for solutions that have pH not affected by glass corrosion mathematically corresponding to a zero surface to volume ratio. The U(t) dependence is due to changes of concentration profiles of elements in the near-surface layers of glasses in contact with water and is most evident within the initial stages of glass corrosion at relatively low temperatures. Numerical examples are given for the nuclear waste borosilicate glass K-26 which is experimentally characterised by an effective diffusion coefficient of caesium $D_{Cs} = 4.5$ 10^{-12} cm^2/day and by a rate of glass hydrolysis in non-saturated groundwater as high as $r_h = 100$ nm/year The changes of U-shaped form need to be accounted when assessing the performance of glasses in contact with water solutions.

INTRODUCTION

Vitrification is currently the most widely used technology for the treatment of high level radioactive wastes throughout the world. Most of the nations that have generated high level wastes are immobilising them either in alkali borosilicate or alkali aluminophosphate glasses; moreover glass has also been used to stabilise a variety of low and intermediate level radioactive and mixed hazardous waste [1]. The most important requirement for a wasteform is good chemical durability where it is desirable for the wasteforms to be highly insoluble in contact with ground waters and in the long-term to minimise potential radionuclide releases to the environment. There is currently a revival of interest in studying glass durability and corrosion processes with many laboratory and field tests as well as studies of archeological glass performed aiming to validate currently available glass corrosion models [2, 3]. We aim in this paper to emphasise the changing character of pH dependence of glass corrosion in conditions when the chemical composition of contacting water is not changed by glass corrosion products e.g. when the glass corrodes in flowing water, or in a large volume of water.
Glass corrosion behaviour is complex and depends on glass composition, environmental conditions (contacting water chemistry, water renewal rate, minerals present, temperature) and time. Fig. 1 demonstrates the pH dependence of a nuclear waste borosilicate glass durability.

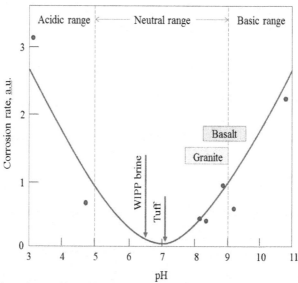

Figure 1. pH dependence of borosilicate glass durability with experimental data taken from [4]. Typical groundwater pH's from candidate repository rock types are indicated, WIPP is Waste Isolation Pilot Plant.

The compositional dependence of glass durability can be estimated using available databases and computer codes such as SciGlass or that of Calculating the Chemical Durability (Hydrolytic Class, Corrosion) of Glasses [5]. Typically an increase of SiO_2 or Al_2O_3 contents increases the glass durability whereas a higher content of alkali diminishes the durability and leads to a higher release of cations into the water.

The time evolution of glass–water interaction depends amongst other on changes induced by corrosion products dissolved in water as they change the chemistry of the water in contact with them. The currently acknowledged progress of glass corrosion in water-confined conditions is given in [1, 2, 6]. The description typically relates to glasses which corrode under confined conditions when the amount of water contacting the glass is small which is likely to be the case under the disposal conditions in a geological repository. Under such conditions the chemistry of water in contact with glass will change with time and that change will affect the corrosion processes. Indeed the ion exchange is characteristic of the initial phase of corrosion and involves replacement of alkali ions in the glass by a hydronium (H_3O^+) ion from the solution. This will cause an ion-selective depletion of the near surface layers of the glass and gives an inverse square root dependence of corrosion rate with exposure time. Glass network dissolution is characteristic of the later phases of corrosion and causes a congruent release of ions into the water solution at a time-independent rate. In closed systems the consumption of protons from the aqueous phase increases the pH and causes a rapid transition to hydrolysis. However, further silica saturation of solution impedes hydrolysis and causes the glass to return to an ion exchange, e.g. diffusion-controlled regime of corrosion [7-9]. The resumption of alteration causes the long term dissolution rate to reaccelerate to a rate that is similar to the initial forward dissolution rate for some glasses. This unexpected and yet poorly understood return to an accelerated rate has been shown to be related to the formation of the Al^{3+} - rich zeolite, analcime, and/or other calcium silicate phases.

However, glass –water interaction evolves in time even for solutions that are kept at constant temperature, and unchanged solution composition and pH. We aim to analyse herewith namely this particular case. These conditions are relevant to corrosion of glasses in flowing ground waters, or when large volumes of waters are in contact with glass. The time evolution in this case is due to changes that occur at the glass-water interface. As a result in non-saturated conditions or in deionised water the transition of one mechanism (ion exchange) to another (hydrolysis) is a function of temperature [9]. It is important to note that these changes in the near surface layers of glass affect also the U-shaped form of corrosion rate pH dependence. We focus our study to pure solutions that are not affected by glass corrosion e.g. when the ratio of water volume to contacting surface is very high.

ANALYSIS OF CORROSION RATE pH DEPENDENCE

Aqueous glass corrosion occurs via two basic mechanisms, namely via diffusion-controlled ion exchange and network hydrolysis (matrix dissolution). The ion exchange is incongruent so that different species are removed from the glass at different rates whereas dissolution occurs congruently, so that glass species are found in the water at the same ratio as in the glass. The diffusion-controlled ion-exchange is actually a process of interdiffusion which is characterised by the normalised release rate [8]:

$$NR_{X,i} = \rho \left(\frac{D_i}{\pi t} \right)^{1/2} = \rho 10^{-mpH} \left[\frac{\kappa D_{0H}}{C_i(0)\pi t} \right]^{1/2} \exp(-\frac{E_{di}}{2RT}) \tag{1}$$

where m=0.5 is the pH power law coefficient , ρ (g/cm^3) is the glass density, D_i (cm^2/day) is the effective interdiffusion coefficient, E_{di} (J/mol) is the activation energy of interdiffusion, R= 8.314 J/ K·mol is absolute gas constant, T is temperature (K), D_{0H} (cm^2/day) is the pre-exponential term in the diffusion coefficient of protons (H_3O^+) in glass, $C_i(0)$ (mol/L) is the cation concentration at the glass boundary and κ (mol/L) is a constant relating the concentration of protons in the glass with concentration of protons in the water, e.g. with the pH. The lower the pH of contacting water the higher the rate of ion exchange $NR_{X,i}$ [8,9]. Note that the normalised leaching rate decreases with time as $t^{-1/2}$.

The dissolution of glass occurs via hydrolysis and is characterised by the normalised dissolution rate [7-10]:

$$NR_H = \rho r_c = \rho k a_{H^+}^{-\eta} \left(1 - \left(\frac{Q}{K} \right)^{\sigma} \right) \exp(-\frac{E_a}{RT}) \tag{2}$$

where r_c is the normalised dissolution rate measured in units of cm/day, k is the intrinsic rate constant (cm/day), a_{H^+} is the hydrogen ion activity in solution e.g. $a_{H^+} = 10^{-pH}$, η is the pH power law coefficient, E_a is the activation energy (J/mol), Q is the dimensionless ion-activity product of the rate controlling reaction, K is the dimensionless equilibrium constant of this reaction and σ is the net reaction order typically taken as σ=1 [10]. Note that for hydrolysis η = 0.5 [7]. The affinity term characterises the decrease of solution aggressiveness with respect to the glass as it becomes increasingly concentrated in dissolved elements and as the ion activity product Q of the reactive species approaches the material solubility product K. The higher the pH of contacting water the higher the dissolution rate NR_H. It is also seen that the normalised dissolution rate is independent of time. Note again that we analyse non-confined water systems which do not change the chemical composition due to glass corrosion and hence they are characterized by constant values of Q. In contrast to the systems analysed herewith, Q changes with time for confined water systems which makes the rate of hydrolysis time dependent.

The normalised corrosion rate of any glass NR_i is given by the sum of the contributions from both incongruent leaching and congruent network dissolution [8, 11]. On varying the pH the $NR_{X,i}$ changes as $10^{-pH/2}$ whereas NR_H changes as $10^{pH/2}$ therefore the pH dependence of NR_i is described by a typical U-shaped curve with minimum in near-neutral water solutions. This explains the well-known experimental result for pH-dependence of glass corrosion in water. However, while the U-shaped curve is characteristic of glasses in contact with water the two contributing corrosion mechanisms have different time dependences. Indeed the corrosion rate can be written as a sum of contributions from ion exchange and hydrolysis

$$NR_i = A_i \cdot 10^{-mpH} + B_i \cdot 10^{\eta pH},$$
(3)

where A_i and B_i are coefficients which can either be determined experimentally or explicitly written based on existing models of elementary processes leading to glass corrosion, and m=η=0.5 for monovalent cations. Indeed the contribution from diffusion-controlled ion exchange is predominant for initial times of corrosion and gradually diminishes at later stages it so that gradually the corrosion becomes mainly due to matrix dissolution via glass hydrolysis. This is true at least in non-saturated conditions. Using relations (1) and (2) we can write for $A_i(t)$ and B_i:

$$A_i = \rho \left[\frac{\kappa D_{0H}}{C_i(0)\pi t} \right]^{1/2} \exp(-\frac{E_{di}}{2RT})$$
(4A)

$$B_i = \rho k \left(1 - \left(\frac{Q}{K}\right)^{\sigma}\right) \exp(-\frac{E_a}{RT})$$
(4B)

Equation (3) gives the generic description of pH dependence of glass corrosion. It demonstrates that the glass corrosion is a two-exponent (two-power) function rather than a simple one-term exponential law which would not have any alternative pH-dependence being either a monotonically-decreasing or increasing function of pH. The minimum of corrosion rate is attained at solution pH given by

$$pH_{i,\min} = \lg[mA_i / \eta B_i]/(m+\eta)$$
(5A)

Accounting that m=η=0.5 this expression can be rewritten as:

$$pH_{i,\min} = pH_{i,\min 0} + \Delta pH_{i,\min}(t),$$
(5B)

where the constant part is given by

$$pH_{i,\min 0} = \frac{E_a}{\ln(10)RT} - \frac{E_{di}}{2\ln(10)RT}$$
(6)

whereas the time-variable part is as follows

$$\Delta pH_{i,\min}(t) = \frac{1}{2}\lg\left(\frac{\tau_{pH}}{t}\right) \tag{7}$$

We have use standard notations $\lg(x) = \log_{10}(x)$ and $\ln(x) = \log_e(x)$. The characteristic time in (7) is given by expression

$$\tau_{pH} = \frac{\kappa D_{0H}}{k^2\left(1 - (Q/K)^\sigma\right)^2 \pi C_i(0)} \tag{8}$$

Note that the activation energies of hydrolysis are typically much higher than those of diffusion-controlled processes, e.g. it is known that over a wide variety of glass compositions E_a is in the range 70–90 kJ/mol [12]. For example, the Russian nuclear waste glass K-26 has an activation energy of hydrolysis $E_a = 68$ kJ/mol which is about twice that of E_{di} [8]. As a result the constant part given by equation (6) is positive $pH_{i,\min 0} > 0$ and is mainly determined by the first term. The variable part of the minimum corrosion rate given by equation (7) decreases slowly with time and becomes negative at times $t > \tau_{pH}$. The older (longer interaction with water) the glass (assuming that it was submitted to aqueous corrosion all that time) the smaller are the changes in the shape of pH dependence of corrosion rates. The characteristic time τ_{pH} when changes in the U-shape of corrosion rate pH dependence become insignificant is given by equation (8). The higher the diffusion coefficient of protons (hydronium) in glass, the lower the cation concentration at the glass boundary the longer are the characteristic times τ_{pH}. The trend is that at very long corrosion times when $t \gg \tau_{pH}$ the U-shaped form remains almost unchanged with the minimum corrosion rate approximately given by the constant part e.g. by $pH_{i,\min 0}$. The constant part of time dependence, as indicated by equation (6), also depends on the type of cation. The corrosion rate is generically not the same for all cations and because of this the pH dependence, and thus the U-shaped form and the pH- minimum of corrosion rate, are different for different cations extracted from glass. This is true both for the constant part of corrosion rate dependence, equation (6), and for the time dependent part given by equation (7).

CHANGES IN U(t) FORM OF pH-DEPENDENCE

Corrosion of silicate glasses including nuclear waste-containing borosilicate glasses involves two major processes – diffusion-controlled ion exchange and glass network hydrolysis. Glass corrosion occurs at low pH preferentially via diffusion-controlled ion exchange whereas at high pH it occurs via hydrolysis. In acidic media below pH = 6 the concentration of protons (or hydronium ions) in the water is high resulting in a high rate of ion exchange. The role of glass network dissolution in this range of pH is insignificant and cation leaching is ion-selective with different leaching curves for different cations. Above pH = 9 the role of ion exchange becomes insignificant due to the high concentration of hydroxyl ions in the water and thus the glass network commences to dissolve rapidly. In such basic media the release of cations becomes congruent as destruction of the glass network results in practically complete dissolution of all glass constituents [13]. Note that the hydrolysis reactions become impeded if solutions become silica-saturated (see Eq. 4B). The pH dependence of glass corrosion is complex when pH changes in the interval $6 < pH < 9$. Moreover attempts to model the pH dependences by simple power laws such as the frequently used

$$NR_i \propto 10^{npH}, \tag{9}$$

are misleading [8]. This will inevitably result in an exponent term n smaller than 0.5 if equation (9) is used instead of the two-exponent generic equation (3). This will in turn result in an underestimation of corrosion caused by hydrolysis at higher pH's.

We give here the results of calculations for nuclear waste glass K-26 which is a borosilicate glass designed to immobilise intermediate level operational nuclear power plant radioactive waste (Fig. 2). Its composition on an oxide basis is (wt%): $43.0 \cdot SiO_2 - 6.5 \cdot B_2O_3 - 3.1 \cdot Al_2O_3 - 13.7 \cdot CaO - 23.9 \cdot Na_2O - 1.9 \cdot Fe_2O_3 - 1.2 \cdot NaCl$ –others [14]. It has been experimentally found that in near surface repository conditions the effective diffusion coefficient $D_{Cs} = 4.5 \ 10^{-12} \ cm^2/day$ and the rate of glass hydrolysis is $r_h = 100$ nm/year [8]. These data enable theoretical calculation of pH-dependence of corrosion rates which was characteristic for the glass in 2003 (Fig. 2).

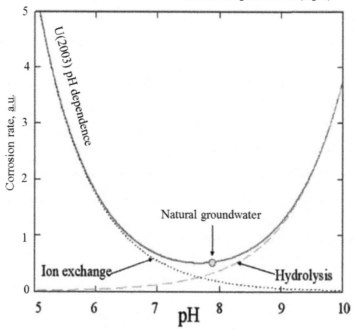

Figure 2. The U-shaped pH dependence of caesium normalised leaching rate and the contributions to this rate from ion exchange and hydrolysis for 16-year old real radioactive glass K-26 in 2003. The experimentally determined rate is indicated by small circle (data from [8]).

The pH minimum of corrosion rate (see solid curve on Fig. 2) is very close to the pH of water in contact with K-26 glass (pH = 7.9 [8]). However, because of the time dependence of ion-exchange rates it is expected that the minimal rates (see equation (5B)) should drift with time to lower values of pH. Fig. 3 shows schematically time changes of the U-shaped form corrosion rates.

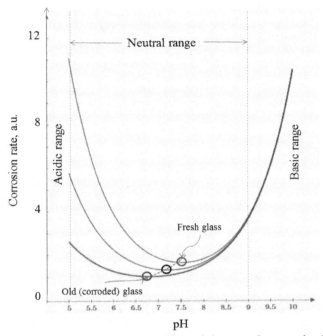

Figure 3. Schematic of time changes of U-shaped form of glass corrosion rates: the older (e.g. a glass which is corroded for longer time) the glass the lower the $pH_{i,min}$ for which the glass corrosion rate reaches its minimum. Corrosion of glass causes the minimal rates to drift with time to lower values of pH.

It is thus seen that the U-shaped form of corrosion rate pH dependence changes with time. A fresh glass in contact with water has ion exchange as its main corrosion mechanism in most natural groundwaters. Consequently, its pH curve has a minimum shifted to higher pH and the U-form is squeezed to the basic range of solutions. Corrosion of glass changes the U-form and a glass which has been in contact with water for a substantial time would have a changed U-shape. Note that standard corrosion testing procedures such as MCC tests involving cleansing of glass powders will therefore slightly shift the pH-dependence of glass durability to higher pH's against that dependence that will really occur. Finally, an old glass with corrosion times exceeding τ_{pH} should have a U-form squeezed to the acidic range of solutions. Note that by old glasses we mean glasses that have been in contact with water for substantial times extending over decades or centuries.

Saturation of solutions with silica also affects the U-shaped form. On approaching the solubility limit the solution aggressiveness with respect to the glass decreases as the ion activity product Q of the reactive species approaches the material solubility product K (see Eq. (4B)). Grambow [15] suggested that only silicon had to be accounted for in the affinity term of Eq. (4B) thereby significantly reducing the number of elements that have to be taken into account. This leads to $B_i \rightarrow 0$ (see Eq. (3)) and thus to $\tau_{pH} \rightarrow \infty$ (see Eq. (8)). The meaning of this change is that the glass corrosion rate pH dependence almost does not change with time.

Although we have used in the above analysis on glass corrosion the model developed in [8, 9] based on Belyustin and Shultz approach [16] an account can be made of changes of cation diffusion profiles caused by dissolution of glasses [17, 18]. Moreover it has been shown that at large times the dominant glass corrosion mechanism in this case is hydrolysis [8, 9]. We should however note that this does not affect the conclusions drawn because of generic character of glass corrosion rate given by equation (3) which accounts for main mechanisms of corrosion.

CONCLUSIONS

This paper has analysed the change of U-shaped form of pH-dependence of glass corrosion rates in water solutions with pH's independent of corrosion process corresponding to conditions when the ratio of contacting glass surface to water volume to is very small e.g. ideally zero. The changing character of pH dependence of glass corrosion caused by varying contributions from two basic mechanisms of glass corrosion – ion exchange and hydrolysis has been analysed. The effects of time changes of the U-shaped form of pH-dependence of glass corrosion due to time variations of these mechanisms have been emphasised. Fresh glasses in contact with water have minimal corrosion rates shifted to higher pH. Glasses which have been in contact with water for a substantial time (older glasses) are characterised by U-shaped form squeezed to the acidic range of solutions. The older the glass (assuming that it was submitted to aqueous corrosion all that time) the smaller are the changes in the shape of pH dependence of corrosion rates. Saturation of solutions leads to almost no changes with time of U-shaped form.

REFERENCES

1. C.M. Jantzen, W.E. Lee, M.I. Ojovan. Radioactive Waste Conditioning, Immobilisation, and Encapsulation Processes and Technologies: Overview and Advances. Chapter 6 in: W.E. Lee, M.I. Ojovan, C.M. Jantzen. *Radioactive waste management and contaminated site clean-up: Processes, technologies and international experience*. p. 171-272, Woodhead, Cambridge (2013).
2. S. Gin, A. Abdelouas, L.J. Criscenti, W.L. Ebert, K. Ferrand, T. Geisler, M.T. Harrison, Y. Inagaki, S. Mitsui, K.T. Mueller, J.C. Marra, C.G. Pantano, E.M. Pierce, J.V. Ryan, J.M. Shoefield, C.I. Steefel, J.D. Vienna. An international initiative on long-term behaviour of high-level nuclear waste glass. *Materials Today*, **16** (6) 243-248 (2013).
3. A. Chroneos, M.J.D. Rushton, C. Jiang, L.H. Toukalas. Nuclear wasteform materials: Atomistic simulation case studies. *J. Nucl. Mater.*, **443**, 29-39 (2013).
4. G. Wicks. Nuclear waste vitrification – The geology connection. *J. Non-Crystalline Solids*, **84** 241-250 (1984).
5. Statistical Calculation and Development of Glass Properties. http://glassproperties.com/ (accessed on 21.05.2014).

6. D. Bacon, E. Pierce. Development of long-term behaviour models for radioactive waste forms. Chapter 14 in: M.I. Ojovan. *Handbook of advanced radioactive waste conditioning technologies*. ISBN 1 84569 626 3. Woodhead, Cambridge, p.433-454 (2011).

7. D.H. Bacon, B.P. McGrail. Waste form release calculations for performance assessment of the Hanford immobilized low-activity waste disposal facility using a parallel, coupled unsaturated flow and reactive transport simulator. *Mat. Res. Soc. Symp. Proc.* **757**, II1.9.1-6 (2003).

8. M.I. Ojovan, R.J. Hand, N.V. Ojovan, W.E. Lee. Corrosion of alkali-borosilicate waste glass K-26 in non-saturated conditions. *J. Nucl. Mater.* **340**, 12-24 (2005).

9. M.I. Ojovan, A.S. Pankov, W.E. Lee. The ion exchange phase in corrosion of nuclear waste glasses. *J. Nucl. Mater.*, **358**, 57-68 (2006).

10. D.H. Bacon, M.I. Ojovan, B.P. McGrail, N.V. Ojovan, I.V. Startceva. Vitrified waste corrosion rates from field experiment and reactive transport modelling. *Proc. ICEM '03: The 9th International Conference on Radioactive Waste Management and Environmental Remediation*, September 21 – 25, 2003, Examination School, Oxford, England, 7p., CD ROM 4509.pdf. (2003).

11. M.I. Ojovan, W.E. Lee. *An Introduction to Nuclear Waste Immobilisation, Second Edition, Elsevier*, 2nd Edition, Amsterdam, 362 p. (2014).

12. D.M. Strachan. Glass dissolution: testing and modeling for long-term behavior *J. Nucl. Mater.*, **298**, 69-77 (2001).

13. W.L. Ebert. The effect of the leachate pH and the ratio of glass surface area to leachant volume on glass reactions. *Phys. Chem. Glasses*, **34** (2) 58-65 (1993).

14. M.I. Ojovan, W.E. Lee, A.S. Barinov, I.V. Startceva, D.H. Bacon, B.P. McGrail, J.D. Vienna. Corrosion of low level vitrified radioactive waste in a loamy soil. *Glass Technol.*, **47** (2), 48-55 (2006).

15. B. Grambow. A general rate equation for nuclear waste corrosion. *Mat. Res. Soc. Symp. Proc.* **44**, 15-27 (1985).

16. A.A. Belyustin, M.M. Shultz. Interdiffusion of cations and concomitant processes in near surface layers of alkali-silicate glasses treated by water solutions. *Physics and Chemistry of Glass*, **9**, 3–27 (1983).

17. Z. Boksay, G. Bouquet, S. Dobos, The kinetics of the formation of leached layers on glass surfaces, *Phys. Chem. Glasses,* **9**, 69-71 (1968).

18. P. Melling, A. Allnatt. Modelling of leaching and corrosion of glass, *J. Non-Cryst. Solids,* **42**, 553-560 (1980).

Mater. Res. Soc. Symp. Proc. Vol. 1744 © 2015 Materials Research Society
DOI: 10.1557/opl.2015.333

Glass Degradation in Performance Assessment Models[1]

William L. Ebert
Argonne National Laboratory,
Argonne, IL, U.S.A

ABSTRACT

The interface with reactive transport models used in performance assessment calculations is described to identify aspects of the glass waste form degradation model important to long-term predictions. These are primarily the conditions that trigger the change from the residual rate to the Stage 3 rate and the values of those rates. Although the processes triggering the change and controlling the Stage 3 rate are not yet understood mechanistically, neither appears related to an intrinsic property of the glass. The sudden and usually significant increase in the glass dissolution rate suggests the processes that trigger the increase are different than the processes controlling glass dissolution prior to that change. Application of a simple expression that was derived for mineral transformation to represent the kinetics of coupled glass dissolution and secondary phase precipitation reactions is shown to be consistent with experimental observations of Stage 3 and useful for modeling long-term glass dissolution in a complex disposal environment.

INTRODUCTION

Performance assessments of geological disposal systems engineered for high-level radioactive waste are conducted to provide confidence that regulations addressing groundwater contamination will be met throughout the regulated service life. The releases of radionuclides and other hazardous constituents to the surrounding biosphere are calculated using reactive-transport models that track contaminant migration through multiple engineered and natural barriers that comprise the disposal system. Waste form degradation models are developed to provide source terms for key contaminants that can be used in transport models throughout the long disposal times. The degradation models quantify the effects of changes in both the waste form surface as it corrodes and the environmental conditions driving the corrosion on the releases of those contaminants. Confidence in the source term values calculated with these models is derived from an understanding of both the processes that control waste form degradation and the influence of environmental conditions on the kinetics of those processes. The modular approach to performance assessment modeling currently being followed in the US has the potential to significantly reduce the degrees of conservatism and empiricism that were necessary in previous waste form models, but adds the need to track solution feed-back and surface alteration effects as waste form corrosion progresses. An approach is being developed by the US Department of Energy for integrating waste form source term models with contaminant transport models for performance assessment. The initial interfaces are being established using the source term model for used oxide fuel, and the insights gained will be applied to source term models

---

[1] This work was under the auspices of the US DOE Fuel Cycle R&D program Materials Recovery and Waste Form Development Campaign. Work at Argonne National Laboratory is supported by the U.S. Department of Energy, Office of Nuclear Energy, under Contract DE-AC02-06CH11357.

being developed for borosilicate glass, ceramic, glass-ceramic, and alloyed waste forms to identify data needs and ensure terms needed to interface with the performance assessment calculations are included. This paper identifies issues to be addressed from the perspective of the glass waste form degradation model and presents a conceptual approach.

The waste package is a container holding several canistered waste forms that is surrounded by backfill material within an excavated disposal chamber. Figure 1 illustrates a breached waste package simplified to identify key processes to be modeled for waste forms in a deep geological repository. The waste form is shown in a breached disposal container that has been partially filled with groundwater that has seeped through the breaches; the waste form canister is considered to be part of the disposal container. Corrosion of the container and canister materials at the breaches generates corrosion products and modifies the chemistry of the water filling the container, which is referred to herein as the in-package solution (IPS). Corrosion of the waste form contacted by the IPS results in dissolution of waste form constituents that further modifies the solution chemistry and generates waste form corrosion products. Sorption on waste form and canister corrosion products, dissolved concentration limits, radioactive decay, and other processes will affect both the contaminant concentrations and the continued corrosion rate of the waste form. The volume of the IPS, area of the waste form contacted by that solution, and amounts of corrosion products are important parameters for both waste form degradation and contaminant transport calculations that can be varied separately. The simplified system shown in Fig. 1 includes some of the processes in a breach canister that affect the IPS composition, which is an important input for applying the glass degradation model. The following sections describe how the glass dissolution rate is taken into account in performance models and aspects of the glass degradation model that will be important to long-term predictions.

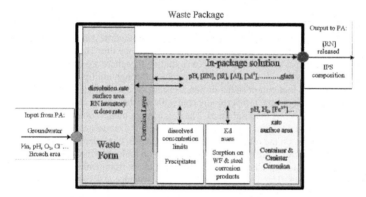

Figure 1. Schematic waste package model showing key processes and interactions.

Role of the Modeled Glass Dissolution Rate in Transport Calculations

In the system shown in Fig. 1, contaminants released from the waste form migrate through the IPS and may be released through a breach in the container. Contaminants can migrate through the backfill material, the disturbed zone in the mined host geology, and then through the natural system to the regulated disposal site boundary. Transport within the breached container and through the surrounding media can be treated mathematically using standard contaminant transport models. A generalized one-dimensional solute-transport equation can be written for each species of interest with terms for dispersion, advection, and reactions occurring either within the waste package or media as

$$\frac{\partial C}{\partial t} = D\frac{\partial^2 C}{\partial x^2} - v_x \frac{\partial C}{\partial x} + \sum_{j=1}^{N_c}\left(\frac{\partial C}{\partial t}\right)_{reaction\ j} \quad , \tag{1}$$

where C is the solution concentration of the species of interest (M L^{-3}), t is time (T), D is the longitudinal dispersion (diffusion) coefficient (L^2 T^{-1}), x is the direction of fluid flow, v_x is the linear velocity of fluid flow in the x direction (L T^{-1}), and reaction j is one of N_c reactions affecting the solute mass of the species of interest. The first two terms on the right-hand side of Eq. 1 are physical processes that govern transport. The last term on the right-hand side represents the summation of all source and sink processes that change the concentration of species in solution by production or loss within the system. These may include chemical transformations such as dissolution, precipitation, oxidation, reduction, radioactive decay, sorption, and ion-exchange reactions. Different reactions and parameter values will be appropriate for the various in-package, near-field, and far-field environments.

The concentrations of two groups of species in the IPS within a breached waste package are of interest to the waste form model: those radioactive and hazardous constituents of the waste form that are important to the performance assessment and those glass constituents that are important to calculating the glass dissolution rate. Source and sink reactions that occur within the breached waste package affect the amounts of contaminants that migrate out of the package and the feedback effects of the IPS on the continued glass dissolution rate. The primary objective of degradation models for high-level radioactive waste forms is to provide the source terms for important dose-contributing radionuclides for use in reactive transport calculations, which are at the heart of the performance assessments. The release of contaminants from the waste form is treated as a source term in the reactive transport model applied to the breached container. This usually requires chemical bonds with the host matrix to be broken by reactions with the solution at sites at or near the surface of the waste form, such that the release rates of most radionuclides are related to the dissolution rate of the glass. Although any process that can be described mathematically can be incorporated into the equation as a source or sink term, most analytical solutions have been derived for chemical transformations expressed as first order reactions. Four reactions of primary interest are glass dissolution, ion-exchange, sorption, and radioactive decay. Radioactive decay occurs by a first order reaction to decrease the concentrations of radionuclides in both the dissolved and sorbed phases. Sorption isotherms are commonly linearized and ion exchange reactions can be modeled as linear sorption processes. The glass dissolution rate expressed in units g m^{-2}d^{-1} can be converted to a first-order rate constant through multiplication by the specific surface area of the waste form expressed in units m^{2}g^{-1}. The general approach is to determine the glass dissolution rate under the conditions of interest using the glass degradation model rate law, and then express that rate as a fractional mass loss based on the geometric dimensions of the waste form. That (constant) rate is used to calculate the change in solution composition that occurs due glass dissolution during the next time step in the transport calculation. This provides the source term for contaminant migration out of the breached waste package and the solution composition used to calculate the glass dissolution rate for the next time step.

Glass Degradation Behavior

The most important uncertainty in projecting the source terms for geologically-disposed high-level radioactive waste glasses is the relevance and magnitude of the sudden increase in the dissolution rates often observed to occur coincidentally with the precipitation of various secondary phases in laboratory tests [1]. Figure 2 shows plots for B concentrations measured in long-term PCTs with several surrogate HLW glasses that did not or did show sudden increases in the dissolution rates [2]. Sudden increases were observed for the majority of HLW and LAW glasses in the database compiled by Jantzen [2], but did not occur in tests with some glasses until after test durations of several years under standard PCT conditions. The PCT is a static test conducted with crushed glass in demineralized water at high glass surface area-to-solution volume ratios (S/V). The solution becomes highly concentrated in dissolved glass constituents after only a few days, which restricts subsequent dissolution to the low rates seen the shorter test durations prior to the increases in Fig. 2b and in all tests in Fig. 2a. The increased rate occurs sooner in tests conducted at higher S/V ratios, but may not occur even after very long test durations for some glasses. The release of B represents the maximum dissolution rate of borosilicate waste glasses. Data points at the longest durations in Fig. 2b indicate almost all of the B has been released from these glasses.

The glass dissolution behavior commonly observed in static and slow flowing systems can be represented as the three reaction stages illustrated in Fig. 3. The shape of the curve describes the dissolution kinetics and includes stages that represent the evolving influences of the corroding surface (e.g., through mass transport limitations) and the solution composition on the glass dissolution rate, primarily the dissolved silica and alumina concentrations and the pH. Dissolution in Stage 1 occurs at a constant rate when the feedback effects of the dilute solutions are negligible and the surface has not been significantly altered. This stage is not discernable in the PCT tests shown in Fig. 2 because the test conditions are designed to study advanced stages of glass corrosion. The reactions have proceeded well into

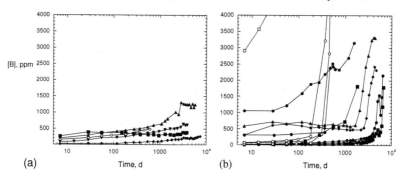

(a) Time, d   (b) Time, d

Figure 2. Plots showing results of PCTs conducted with various surrogate HLW glasses that (a) did not show a sudden increase in the dissolution rate and (b) tests that did. Filled symbols for tests conducted at about 2000 m^{-1} and open symbols for tests conducted at about 20,000 m^{-1}.

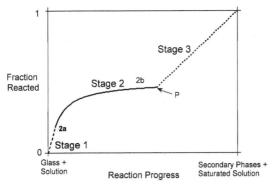

Figure 3. Schematic representation of glass dissolution and reaction progress (not to scale).

Stage 2 by the first data points. Dissolution in Stage 2 slows as the silica concentration approaches a quasi-saturation limit with respect to the chemically and physically altered glass surface. The dissolution rate approaches a minimum limiting rate that is referred to as the residual rate and is labeled as Point 2b in Figure 3. This is seen as the slowly increasing concentrations in Fig. 2. Both the solution and surface layers likely affect the residual rate. The dissolution rate increases when certain secondary phases precipitate (at Point P in Fig. 3) to initiate Stage 3. Long test durations at high S/V ratios are usually required for Stage 3 behavior to be observed experimentally. As seen in Fig. 2b, the increased release is sudden and significant, and continues until the glass is completely dissolved.

The experimentally observed glass corrosion behavior represented in Figs. 2a and 2b indicate the fraction of glass that dissolves prior to the increase is very small and can be represented using a constant source term in performance assessments. In terms of Fig. 3, two major challenges to modeling long-term waste glass dissolution rates and radionuclide source terms in disposal systems are determining (1) if, how, and when Stage 3 is triggered (Point P) and (2) the appropriate rate law for Stage 3 dissolution under the prevailing environmental conditions. Although the mechanisms triggering and controlling dissolution in Stage 3 remain to be determined, the Stage 3 rate is usually observed to be constant, as seen in Fig. 2b and depicted in Fig. 3, and not related to the glass composition in a simple manner. A possible reason why the rates in Stages 2b and 3 are nearly constant is discussed below. A third major challenge is triggering the change in rate within the glass degradation model during the performance assessments.

Glass Degradation Model

Processes contributing to glass corrosion and transformation include dealkalization and ion exchange reactions with (primarily) alkali metals and alkaline earth elements in the glass, hydrolysis reactions, and water diffusion into alteration layers formed by these reactions and the underlying glass structure [3]. Because many of these processes proceed simultaneously, modeling glass corrosion must take into account the coupling of these processes and changes to the suite of reactions that occur as corrosion progresses. This includes the coupled evolution of the solution and alteration layers on the glass surface. The development of waste glass dissolution models has been closely linked with those developed for mineral dissolution. The predominant model used for waste glass dissolution for the past 30 years is based on the reaction affinity model formulated by Aagaard and Helgeson [4] for the dissolution of aluminosilicate minerals, which is

$$rate_{net} = rate_{forward} \bullet \left[1 - \exp\left(\frac{\Delta G_r}{RT} \right) \right] = rate_{forward} \bullet \left[1 - \left(\frac{Q}{K} \right) \right], \tag{3}$$

as applied to glass dissolution by Grambow [5]. The second form expresses the free energy as $\Delta G_r = RT \ln (Q/K)$. This rate law has separate terms that represent the environmental effects on the kinetics of the dissolution reactions ($rate_{forward}$) and attenuation factors imposed by the thermodynamic affinity for those reactions (the term in brackets) [6]. An empirical expression commonly used to express the dependence of forward dissolution rates of glasses on glass composition, solution composition, and temperature is

$$rate_{forward} = A\, k_0 \bullet \Pi_i (a_i)^{\eta_i} \bullet \exp\left(-E_a \big/ RT \right), \tag{4}$$

where A is the surface area (L^2), k_0 is the intrinsic rate coefficient ($M\,L^{-2}\,T^{-1}$), a_i is the activity of solute i (unitless), η quantifies the dependence of the rate on solute i (unitless), E_a is the activation energy ($kJ\,mol^{-1}$), R is the gas constant ($kJ\,mol^{-1}\,K^{-1}$), and T is the temperature (K). This equation is consistent with the analytical pH and temperature dependencies and parameter values of dissolution rates measured for a wide range of surrogate borosilicate waste glasses. It provides the overall rate for all reactions contributing to glass dissolution and describes stoichiometric dissolution. The effects of other processes (e.g., water diffusion and ion exchange) are represented implicitly by the values of terms in Eqs. 3 and 4 or lumped in the parameter values derived from experiments. Mass transport is represented explicitly in some models and implicitly in others. For example, mass transport restrictions may attenuate the $rate_{forward}$ term through the activity of a reactant at a reaction site that is lumped with the k_0 term.

The reaction affinity term in Eq. 3 quantifies the degree of disequilibrium in the system relative to the thermodynamically most stable state of that system (i.e., the free energy). The analytical form of the affinity term is empirical and the simple linear form used in Eq. 3 may not be applicable to all systems and may not hold over the full range of disequilibrium in an evolving system as the rate-controlling mechanism changes or competing reactions occur [7, 8]. This is probably the case for waste glass corrosion, where changes in the composition and structure of the reacting surface and the generation of a wide range of secondary phases can introduce significant changes in the reacting system that affect both the kinetic and thermodynamic terms in the rate law. It is likely that the sudden change in the dissolution rate triggering Stage 3 is due to a change in the form of the affinity term.

In the case of mineral dissolution, the thermodynamically stable state may be a saturated solution in contact with the same mineral or a secondary mineral. Because glass is thermodynamically unstable and cannot equilibrate with the solution, the durability of glass is due solely to its slow dissolution kinetics. Grambow treated the solution conditions for which glass dissolution is immeasurably slow as a pseudo-equilibrium state and used the solubility of amorphous silica to represent the "stability constant" for the glass [5]. In practice, the value of that stability constant has been selected to match experimental results [9]. The effect of secondary phase formation on the glass dissolution rate has been attributed to the greater thermodynamic stabilities of those secondary phases relative to amorphous silica [10]. It was hypothesized that the formation of a more stable secondary phase increased the reaction affinity by consuming orthosilicic acid (H_4SiO_4) from the solution, but with no change to the stability constant or analytical form of the affinity term [11, 12, 13]. As will be described below, this represents application of the concept of partial equilibrium to the system, wherein secondary phases are assumed to precipitate instantaneously and remain in equilibrium with the solution. That approach may not be appropriate for modeling glass dissolution or Stage 3 behavior.

In a reacting system, the dissolution and precipitation reactions are coupled by the transfer of species from the dissolving phase (glass) to growing secondary phase through the solution. A reaction path model describing mass transfer between primary and secondary minerals was developed by Helgeson by applying thermodynamic principles to geochemical processes [14, 15]. Linear differential equations representing reversible and irreversible reactions were used to track the redistribution of species due to the appearance and disappearance of stable and metastable phases. As summarized by Zhu and Lu [16], this so-called classical partial equilibration approach is based on the following assumptions: (1) the irreversible dissolution of the primary material is the rate-limiting reaction that drives all processes in the system; (2) all secondary phases precipitate instantaneously; (3) the aqueous solution is either under-saturated or at equilibrium with all secondary phases at all times; (4) the fluid chemistry obeys the phase rule such that when three phases are present, the fluid must follow a linear phase boundary until one of the solid phases is no longer present; and (5) secondary phases dissolve and precipitate in a paragenetic sequence as the solution chemistry evolves. Such a system is said to be in partial equilibrium, wherein the aqueous species are in equilibrium with the precipitating secondary phases but not in equilibrium with the dissolving primary phase; mass transfer is driven solely by the irreversible dissolution of the primary phase. Zhu has referred to the precipitation of secondary phase in this model as providing a positive feedback loop, wherein the precipitation of more secondary phase causes more of the primary phase to dissolve, and vice versa [17]. In this case, the precipitation rates of secondary phases are limited only by the dissolution rate of the primary phase and the coupling between the dissolution and precipitation reactions is weak.

A stronger coupling can occur between the dissolution and precipitation kinetics when the secondary phases are not in equilibrium with the solution or do not precipitate instantaneously. Lasaga demonstrated the importance of the relative values of rate constants for the dissolution and precipitation reactions on the deviation of a system from equilibrium [18]. Using an analytical approach, he predicted the persistence of metastable precipitates due to the sluggishness of most precipitation reactions relative to the dissolution reaction and that the gross differences in reactive surface areas can lead to significant deviations from the partial equilibrium model. Nagy et al. [19] showed the coupled reactions of a dissolving primary phase (glass) and precipitating secondary phase near steady state occur at the rate

$$rate_{coupled} = \frac{rate_{forward}^{(1)} \; rate_{forward}^{(2)}}{rate_{forward}^{(1)} + rate_{forward}^{(2)}} \left[1 - \exp \left(\frac{\Delta G^{(1)} + \Delta G^{(2)}}{RT} \right) \right],$$ (5)

where the superscripts (1) and (2) represent the dissolution and precipitation reactions, respectively. Now consider application of Eq. 5 under the conditions assumed in the partial equilibrium model. The secondary phases are assumed to be in equilibrium with the solution at all times, such that $\Delta G^{(2)} \equiv 0$ and the precipitation of secondary phases occurs much faster than dissolution. When $rate^{(2)}_{forward} >> rate^{(1)}_{forward}$ and $\Delta G^{(2)} = 0$, Eq. 5 reduces to Eq. 3 and the formulation used by Grambow. Under conditions that the partial equilibration model holds, the effect of secondary phase formation on the glass dissolution rate occurs through the effect of secondary phase precipitation on the solution composition.

At the other extreme, when the precipitation rate is much lower than the forward dissolution rate of the primary phase, $rate^{(2)}_{forward} << rate^{(1)}_{forward}$, Eq. 5 reduces to

$$rate_{coupled} = rate^{(2)}_{forward} \left[1 - \exp\left(\frac{\Delta G^{(1)} + \Delta G^{(2)}}{RT} \right) \right]. \tag{6}$$

The steady state condition established by the coupled dissolution and precipitation reactions fixes the sum of the free energy terms at a non-zero value that prevents progress towards equilibrium and maintains a constant rate. That is, reaction progress that decreases $|\Delta G^{(2)}|$ will increase $|\Delta G^{(1)}|$ equally and vice versa, such that the sum $\Delta G^{(coupled)} = \Delta G^{(1)} + \Delta G^{(2)}$ remains constant. The kinetic limitations for dissolution and precipitation affect the thermodynamic drivers for both. The reaction affinity for the system becomes "arrested" at a constant value and the dissolution rate (e.g., of the glass) remains constant for as long as the steady state between the dissolving and precipitating phases can be maintained. This could be until a more stable secondary phase forms or until the dissolving phase is almost completely consumed. A rate law with the form of Eq. 6 would maintain the constant Stage 3 rate that is observed experimentally whereas Eq. 3 would not. The precipitation rate of the secondary phase will only control the steady-state rate if it is much slower than the dissolution rate Otherwise, the rate will be some combination of the dissolution and precipitation rates. For example, the slow precipitation rate of clay has been purported to limit feldspar reaction rate in an aquifer [17, 20]. Analogously, the precipitation of clay could be a contributing process that maintains the residual rate observed during glass dissolution.

An important aspect of Eq. 7 is that the affinity term depends on the free energies of both the dissolution and precipitation reactions regardless of the kinetic term. The affinity term in the Grambow model represents the disequilibrium between the solution (Q) and equilibrium product $K^{(1)}$ for an altered glass surface similar to amorphous silica. The free energy term for the coupled dissolution and precipitation reactions in Eq. 5 includes the disequilibrium between the solution and **the secondary phase**. Although dissolved silica (orthosilicic acid) has been shown to dominate the reaction affinity for glass dissolution, additional species will affect the reaction affinity for secondary phase precipitation. A system in which parallel reactions occur to form different secondary phases will require a more complicated analytical form of the affinity term, but the same principles apply based on the relative rates of dissolution and precipitation.

Secondary phases provide stable sinks accepting the transfer of Al, Si, alkali metals, and other components from the glass and the coupled reaction may decrease or increase the glass dissolution rate depending on the precipitation rates of those secondary phases. Slowly precipitating phases (e.g., phyllosilicates and clay) can form without triggering an increase in the glass dissolution rate, and may instead trigger a decrease based on negative feedback. The more rapid precipitation rates of zeolites (analcime, Na-chabazite, phillipsite) are likely to trigger an increase and provide positive feedback. The compositional differences between the glass and suite of alteration phases may result in one glass constituent limiting the transformation rate. The Al concentration in solution is commonly seen to decrease when secondary phases triggering Stage 3 precipitate. The "arrested" reaction affinity predicted by the coupled rate expression in Eq. 6 is consistent with the constant residual rates and Stage 3 rates that are observed experimentally.

Finally, it is important to realize that the PCTs represented in Figs. 2a and 2b were closed systems in which only species released as the glasses dissolved were available to generate secondary phases. In actual disposal systems, water contacting waste glasses will include other species present in the groundwater and from corrosion of engineering components and backfill materials that will complicate the evolution of the IPS composition over time. Therefore, the objective of the conceptual model described herein is not to reproduce test results such as those shown in Figs. 2a and 2b per se; rather, the objective is to use those test results to identify conditions triggering Stage 3 behavior and the underlying mechanisms to develop a degradation model that includes the effects of the IPS composition on the glass dissolution kinetics and can be used for performance assessment calculations.

CONCLUSIONS

The same deviations from the partial equilibrium model seen in mineral transformation reactions due to the relative rates for dissolution and precipitation reactions can limit its application to glass corrosion. Perhaps most importantly, the reaction(s) dominating the affinity term may change when secondary phases form. A conceptual model is proposed in which (1) the kinetic term of the rate law represents the coupled rates for glass dissolution and secondary phase precipitation rather than for glass dissolution alone and (2) the reaction affinity term represents the free energy due to disequilibrium between the solution and the assemblage of secondary phases present in the system, rather than disequilibrium of the glass alone (as represented by amorphous silica). Important aspects of the proposed model are that the mass transfer restrictions that establish the steady state conditions linking the dissolution kinetics with precipitation kinetics, the analytical form reaction affinity term, and contributing species can change significantly when increasingly stable phases form. The newly formed phases lower the free energy of the system and thermodynamic driver for glass corrosion and change the coupled kinetics.

This conceptual model is consistent with experimentally observed glass dissolution behavior of a constant very low residual dissolution rate that may suddenly increase to a much high nearly constant Stage 3 dissolution rate. Secondary phase formation can provide either negative or positive feedback to the glass dissolution rate. The very slow kinetics of clay formation may control the low residual rate, whereas the more rapid formation of zeolites may provide an alternative reaction path resulting in the higher Stage 3 rate. The rate expression for the coupled dissolution and precipitation reactions represents steady-state rates consistent with the nearly constant residual and Stage 3 rates that are measured in laboratory experiments. Both the residual and Stage 3 rates can be readily incorporated into a solution-based glass waste form degradation model for use in performance assessment calculations. However, neither the residual rate nor the Stage 3 rate represents an intrinsic property of the glass and the relationships between both rates and the IPS must be determined experimentally.

REFERENCES

1. Fournier, M., Frugier, P., and Gin, S. (2014). "Resumption of nuclear glass alteration : State of the art." *Journal of Nuclear Materials, 448,* 348-363.
2. Jantzen, C.M. (2013). *Letter Report on SRNL Modeling Accelerated Leach Testing of Glass (ALTGLASS).* SRNL-L3100-2013-00177; FCRD-SWF-2013-000339, Rev. 0.
3. Gin, S. et al. (2013). "An international initiative on long-term behavior of high-level nuclear waste glass." *Materials Today* 16(6).
4. Aagaard, P. and Helgeson, H.C. (1982). "Thermodynamics and Kinetic Constraints on Reaction Rates among Minerals and Aqueous Solutions. I. Theoretical Considerations," *American Journal of Science, 282,* 237-285.
5. Grambow, B. (1987). *Nuclear waste glass dissolution: Mechanism, Model, and Application.* Swedish Nuclear Fuel and Waste Management report JSS-TR-87-02.
6. Lasaga, A.C. (1981). "Rate Laws of Chemical Reactions," in *Reviews in Mineralogy, Vol. 8: Kinetics of Geochemical Processes,* ed. A.C. Lasaga and R.J. Kirkpatrick, pp. 1-68.
7. Burch, T.E., Nagy, K.L., and Lasaga, A.C. (1993). "Free Energy Dependence of albite dissolution kinetics at 80 °C and pH 8.8." *Chemical Geology, 105,* 137-162.
8. Maher, K., Steefel, C.I., White, A.F., and Stonestrom, D.A. (2009). "The role of reaction affinity and secondary minerals in regulating chemical weathering rates at the Santa Cruz Soil Chronosequence, California." *Geochimica et Cosmochimica Acta, 73,* 2804–2831.
9. Grambow, B, and Strachan, D.M. (1988). *A comparison of the performance of nuclear waste glasses by modeling.* Pacific Northwest National Laboratory report PNL-6698.
10. Van Iseghem, P. and Grambow, B. (1988). "The Long-Term Corrosion and Modeling of Two Simulated Belgian Reference High-Level Waste Glasses." Scientific Basis for Nuclear Waste Management XI. Material Research Society Symposium Proceedings, 112, 631-639.
11. Strachan, D.M. and Croak, T.L. (2000). "Compositional effects on long-term dissolution of borosilicate glass." *Journal of Non-Crystalline Solids, 272,* 22-33.
12. Grambow, B. and Müller, R. (2001). "First-order dissolution rate law and the role of surface layers in glass performance assessment." *Journal of Nuclear Materials, 298,* 112-124.
13. Strachan, D.M. and Neeway, J. (2014). "Effects of alteration product precipitation on glass dissolution." *Applied Geochemistry 45,* 144-157.
14. Helgeson, H.C. (1968). "Evaluation of irreversible reactions in geochemical processes involving minerals and aqueous solution—I. Thermodynamic relations." *Geochimica et Cosmochimica Acta 32,* 853-877.
15. Helgeson H. C. (1979). "Mass transfer among minerals and hydrothermal solutions." In Geochemistry of Hydrothermal Ore Deposits (ed. H. L. Barnes). John Wiley & Sons, New York, pp. 568–610.
16. Zhu, C. and Lu, P. (2009). "Alkali feldspar dissolution and secondary mineral precipitation in batch systems: 2. Saturation states of product minerals and reaction paths." *Geochimica et Cosmochimica Acta, 73,* 3171-3200.
17. Zhu, C. (2009). "Geochemical modeling of reaction paths and geochemical reaction networks." in Reviews in Mineralogy & Geochemistry, Vol. 70, pp. 533-569.
18. Lasaga, A. (1998). Kinetic Theory in the Earth Sciences, Princeton University Press, Princeton, NJ. (See section 1.12)
19. Nagy, K.L., Blum, A.E., and Lasaga, A.C. (1991). "Dissolution and precipitation kinetics of kaolinite at 90 °C and pH 3: The dependence on solution saturation state." *American Journal of Science, 291,* 649-686.
20. Zhu, C., Lu, P., Zheng, Z., and Ganor, J. (2010). "Coupled alkali feldspar dissolution and secondary mineral precipitation in batch systems: 4. Numerical modeling of kinetic reaction paths." *Geochimica et Cosmochimica Acta, 74,* 3963-3983.

Mater. Res. Soc. Symp. Proc. Vol. 1744 © 2015 Materials Research Society
DOI: 10.1557/opl.2015.469

Hierarchical Modeling of HLW Glass-Gel-Solution Systems for Stage 3 Glass Degradation

Carol M. Jantzen[1] and Charles L. Crawford[1]
[1]Environmental & Chemical Process Technology
Savannah River National Laboratory
Aiken, SC 29898, U.S.A.

ABSTRACT

The necessity to a priori predict the durability of high level nuclear waste (HLW) glasses on extended time scales has led to a variety of modeling approaches based primarily on solution (leachate) concentrations. The glass composition and structure control the leachate and the gel compositions which in turn control what reaction products form: the leached layer is a hydrogel and reacts with the solution (leachate) to form secondary phases some of which cause accelerated glass dissolution which is undesirable. Glasses with molar excess alkali that is not bound to glass forming $(Al,Fe,B)O_4$ structural groups in the glass resume accelerated leaching. The hydrogels of the glasses that resume accelerated leaching at long times contain excess alkali and the leachates contain excess strong base, $[SB]_{ex}$. The $[SB]_{ex}$ further accelerates aluminosilicate gel aging into analcime with time. Glasses with no excess molar structural alkali do not resume accelerated leaching: the glass generates weak acids, $[WA]$, in the leachate favoring hydrogel aging into clays. These data indicate that the gel layer transforms to secondary phases in situ in response to interactions with the chemistry of a continuously evolving leachate.

INTRODUCTION

Current theories of glass dissolution suggest that all glasses typically undergo an initial rapid rate of dissolution denoted as the "forward rate" (Figure 1-Stage I). However, as the contact time between the glass and the leachant lengthens some glasses come to "steady state" equilibrium and corrode at a "steady state" or "residual rate (Figure 1-Stage II) rate while other glasses undergo a disequilibrium reaction with the leachant solution that causes a sudden change in the solution pH or the silica activity in solution [1]. The "resumption to an accelerated rate" (Figure 1-Stage III) after achieving "steady state" or "residual" dissolution is undesirable as it causes the glass to return to a rapid dissolution characteristic of initial dissolution and confounds long term durability modeling of nuclear waste glasses.

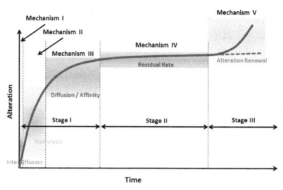

Figure 1. Generally accepted glass dissolution mechanisms.[2]

The initial rate of dissolution is often referred to as Stage I but encompasses Mechanisms I, II, and III (see Figure 1 labelling) because multiple mechanisms are operative including regimes that are inter-diffusion controlled, hydrolysis controlled, and diffusion or affinity controlled which leads to a dissolution rate drop.[3]

In the region of diffusion/affinity control, which starts in Stage I and continues into Stage II, the concentration of the glass components increase in solution, the solution approaches saturation with respect to some rate-limiting alteration phase(s) but is not at final or steady state saturation. In other words, the dissolution rate slows down as the chemical potential difference between the glass and contacting solution decreases. In the case of a glass or other solid phases, this effect can be expressed mathematically as the chemical affinity of a reaction.[4] The dissolution at this stage may be controlled by an interfacial dissolution–reprecipitation mechanism as noted for silicate minerals, i.e. the gel is both forming and condensing.

A recent mechanistic model of glass durability, including the slowing of the dissolution rate (Stage II) due to affinity and/or surface layer effects, was presented by Grambow and Muller [5] and is referred to as the GM2001 model. The GM2001 model combines the effect of glass hydration by water diffusion with ion exchange and affinity-controlled glass network corrosion. The slowing of dissolution due to the effect of a growing surface gel layer is represented by a mass transfer resistance for silica by this layer.

At the interface between the glass and the gel layer, a compositionally different "gel layer" is assumed to be hydrated glass that allows diffusion of H_2O in and boron and alkali atoms out. This layer is believed [3,6,7] to be a diffusion barrier for reactive aqueous species that allows the glass to dissolve at a rate which is several orders of magnitude below the initial (forward rate). A 2003 modification of the GM2001 model, known as the GM2003 model [3], treats silica dissolution and silica diffusion through the gel separately from water diffusion and boundary conditions are specified at the gel/diffusion layer and the gel/solution interfaces. Recently, the GRAAL (Glass Reactivity with Allowance for the Alteration Layer) model [6,7] has been proposed which is dependent on the composition and the passivating nature of the gel layer, called the Passivating Reactive Interphase (PRI). The leached layer has been found experimentally to be zoned (~5-7 zones) [3,8] and the GRAAL model assigns various mechanisms to different zones within the PRI. Earlier studies [9] using Analytic Electron Microscopy (AEM), Scanning Electron Microscopy (SEM), X-Ray Diffraction (XRD), Infrared Spectroscopy (IR), Electron Microprobe Analysis (EMPA), and Secondary-ion Mass Spectroscopy (SIMS) combined with ion microprobe imaging (IMI), had resolved 6 layers. In these earlier studies, a clear trend was observed in the degree of structural order and crystallinity for the sequence of bands: bands closer to the unreacted glass core were amorphous while the outer layers were more crystalline.[9] A more detailed summary of some of the current analytic and modeling approaches can be found in Reference 2.

In the GM models and the GRAAL model, the on-set of Stage II is assumed to be due to formation of an altered surface layer having solubility properties similar to those of amorphous silica. In other words, a dynamic equilibrium is assumed between the hydrolysis and condensation of Si-O bonds at the altered surface whereas such equilibrium cannot exist for the fresh glass surface due to presence of other species. This follows a suggestion by Oelkers [10] who differentiates the properties of the native glass and the altered glass surface. The disequilibrium between the solution and this altered surface, which is in reality an alteration phase, establishes the thermodynamic reaction affinity for the system, including the reactions leading to glass dissolution.

During Stage III, the classic interpretation of durability presented in the previous paragraphs is that the impacts of a rate-limiting alteration phase, in this case an aged gel layer, is being sensed by the overall system. Therefore, the undesireable "resumption to an accelerated rate" appears to be driven by whether the gel layer that forms on the glass surface during dissolution ages into clay or zeolite minerals. Zeolite mineral assemblages form at higher solution pH and preferentially appear associated with Al^{3+} rich glasses. Gels that aged into clay mineral assemblages continued leaching at constant long-term rates.

However, analytic interrogation of the gel layer chemistry for some HLW glasses [11,12] has indicated that glasses that resumed leaching had gel compositions of ~$Na_{0.84}Al_{0.98}Si_{2.0}O_{5.89} \bullet xH_2O$, which is hydrated analcime. Glasses that dissolved at a constant long-term rate were found to have gel compositions of ~$Fe_2Si_2O_5(OH)_4 \bullet 2H_2O$ (hisingerite) with excess amorphous SiO_2. Hisingerite possesses the same local structure as nontronite clay and is a poorly-crystallized precursor of ferric smectite clays.[11] Glasses with excess molar alkali, $(Na,Li,K)_2O$, over the sum of molar $(Na,Li,K)AlO_4 + (Na,Li,K)FeO_4 + (Na,Li,K)BO_4$ resumed leaching and contained excess alkali in the gel and excess strong base, $[SB]_{ex}$, in the leachate.[12] Glasses without excess alkali did not resume leaching and the leachate continued to generate excess weak acids, [WA], in solution with reaction progress favoring gel aging into clays. Identification of the rate limiting precursor complexes appears to be dependent on leachate [SB] and [WA] interactions with the hydrogel. [11,12] In glasses without $[SB]_{ex}$ the mechanisms described above, i.e. that a sufficient fraction of silica is retained in the gel, are reinforced because the leachate is either neutral (buffered with respect to [SB] and [WA]) or have excess weak acids $[WA]_{ex}$. It is shown in this paper that the interactions of the gel with the leachate $[SB]_{ex}$ or $[WA]_{ex}$ controls the composition and the passivating nature of the gel layer.

ALTGLASS DATABASE DEVELOPMENT

A long term durability database, Accelerated Leach Testing of GLASS (ALTGLASS), was developed in 2013. ALTGLASS Version 1.0 contained 213 glasses of which 74 are high-level waste (HLW) glasses. From the 74 HLW glasses populations of 12 {glass-gel-solution} systems were studied [11,12]: some known to resume leaching at an accelerated rate and some that continued at a steady state rate. New accelerated leach data has been generated on five HLW glass standards often used in HLW Performance Assessments (PAs) such as the Defense Waste Processing Facility (DWPF) Environmental Assessment (EA) glass [13] and the DWPF Waste Compliance Plan (WCP) glasses.[14] The EA and WCP glass data has recently been incorporated into ALTGLASS Version 2.1 [15] and this data is assessed in this study and compared to the French SON68 glass (a nonradioactive surrogate of the French R7T7 HLW glass) dissolution and the Advanced Fuel Cycle Initiative (AFCI) glass dissolution. The SON68 and the AFCI glass are common to the HLW population of 12 glasses previously studied [11,12] and to the current study.

ALTGLASS is partially partitioned into hierarchical "parent-child" categories, i.e. the parent group is nuclear waste glasses, the children are HLW and low activity waste (LAW) glasses (Figure 2). The grandchildren in the hierarchical database are those HLW populations that continue leaching at a steady state rate and those that resume accelerated leaching at long times. Great-grandchildren "types" were defined for ~12 HLW glasses in the ALTGLASS database by Muller, et al.[16], Ribet et.al.[17], and Jantzen et.al.[11,12]. The LAW grandchildren populations are being defined by researchers at Argonne National Laboratory although a different classification criteria is being used in the LAW study than in the Muller, et al. and Ribet et.al [16,17] study. The LAW glasses will not be discussed but it should be noted that a high percentage (~82%) of the LAW glasses "return to the accelerated rate" as they are high alkali glass compositions.

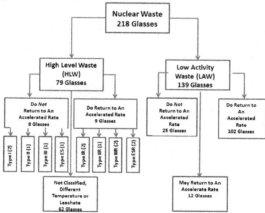

Figure 2. Representation of ALTGLASS as a Hierarchical Database for Modeling.

EXPERIMENTAL

The ALTGLASS database is composed of PCT (ASTM C1285) tests at the reference mesh (-100 to +200) sample size and using various accelerated parameters such as more grams of glass per leachant volume and longtime durations. The PCT-B test was recommended as an indicator of long-term glass dissolution behavior at surface area to volume ratios of 20,000m^{-1} and 90°C.[18]
    Almost all the glasses have been leached in deionized water and not groundwater in closed system conditions at 90° +/- 2° C. For this study on the WCP and EA glass standards duplicate tests in stainless steel vessels via the PCT-A (for 7 day test) and PCT-B (longer duration tests) procedure [19] were performed. Glass powders were prepared using a tungsten blade grinder and brass sieves and the sieved powders were washed with ultrapure water and absolute ethanol according to the PCT procedure. Glass to leachant ratios varied from 1g:10mL, 2g:10mL, 5g:10mL and 6g:6mL, respectively for the 7, 28, 60, and 90 day tests. These ratios resulted in nominal surface area to volume (SA/V) ratios of 2,000m^{-1}, 4,000m^{-1}, 10,000m^{-1} and 20,000m^{-1}, respectively for the increasing test durations. For the SON68 and AFCI glasses all the tests used the 100/200 mesh at 1:1 g/mL ratio which gave a nominal surface area to volume (SA/V) ratio of 20,000m^{-1}. Test durations were 7 to 600 days and the data is reported in the ALTGLASS database.[15] Leachates were filtered using a 0.45 micron nylon syringe filter and diluted using ~ 1 vol% ultrapure nitric acid before analysis by Inductively Coupled Atomic Emission Spectroscopy (ICP-AES).
    The comparison of the EA, WCP, SON68 and AFCI glasses were chosen because this suite of seven HLW glasses, which includes U.S. and French standard glasses, flexes a wide range of HLW glass compositions based on a substitution of glass forming species, i.e. Al$_2$O$_3$ vs. Fe$_2$O$_3$ and Al$_2$O$_3$ vs. B$_2$O$_3$ (Figure 3) which have previously been shown to be important to the long term durability of HLW glasses.[11,12,20]

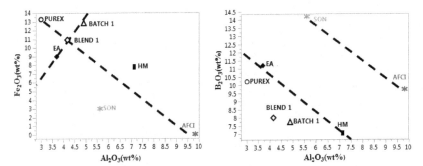

Figure 3. Compositional envelope covered by the seven HLW glasses in this study.

DISCUSSION

    The B released to the leachate is used to define the reaction progress as B is released to the leachate faster than any other element. Reaction progress is therefore defined as:

$$(\text{Reaction Progress})_B \equiv \frac{(C_B)_{leachate}(Volume)_{leachate}}{(f_B)_{glass}(Mass)_{glass}} \tag{1}$$

where

Reaction Progress	=	% of glass leached
C_B(leachate)	=	concentration of element "B" in the leachate with waste form, g_i/L, and
f_B	=	mass fraction of element "B" in the unleached waste form (g_i/g_{glass}).
Volume	=	volume of leachate (L)
Mass	=	mass of glass (g_{glass})

Based on either B release or reaction progress, the EA glass and WCP Purex glass were found to be glasses that return to an accelerated rate of dissolution at long times as was the AFCI glass (Figure 4). The remaining three WCP glasses (Blend 1, Batch 1, and HM) and the SON68 are glasses that do not return to an accelerated rate of dissolution even at long times Figure 4).

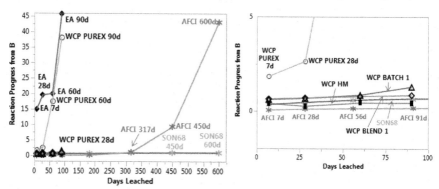

Figure 4. Reaction Progress for the seven HLW glasses based on boron in the leachate shown at full scale (left) and the detail shown as an inset (right).

Buffered vs. Non-Buffered Leachates

During static leaching in deionized water the glass chemistry causes a strong base/weak acid or strong base minus weak acid, i.e. [SB]-[WA], equilibrium to occur in the leachant.[21,22] The silicic and boric acids are weak acids that form from the hydrolysis of the SiO_2 and B_2O_3 in the glass. The strong bases (OH⁻) form from the hydrolysis of the alkali and alkaline earth species (Li_2O, Na_2O, K_2O, CaO, BaO, etc.) in the glass. The dissolution of an alkali species in the glass via hydrolysis releases one mole of OH⁻ for each mole of elemental alkali or ½ mole of an alkaline earth, e.g., each Na^+ releases one mole of OH⁻ or a Ca^{2+} releases two moles of OH⁻ (Equation 2) where OH⁻ is a strong base and is written as a molar concentration [SB].

$$M_2O + SiO_2 + B_2O_3 + 6H_2O \rightarrow 2M^+ + \underbrace{2OH^-}_{strong\,base} + \underbrace{H_4SiO_4}_{weak\,acid} + \underbrace{2H_3BO_3}_{weak\,acid} \tag{2}$$

Conversely, Si and B are present in the leachate solution as weak acids (H_4SiO_4 and H_3BO_3) as represented in Equation 2 or weak acid salts ($MHSiO_3$ or MH_2BO_3). Each mole of weak acid, written as a molar concentration [WA] buffers one mole of strong base via interactions between the hydrolysis reaction(s) and the neutralization reactions of the weak acids at various pH values.

The concentrations of the leachates are expressed in millimoles/L (mM/L) of each element in solution. The molar strong base [SB] generated by the hydrolysis of the alkali oxides in the leachate solution can be approximated by summing the molar alkalis measured in solution in mM/L.[21,22] The alkaline earths can be added to the [SB] as deemed necessary by the glass or leachate composition. The molar weak acids [WA] generated by the hydrolysis of the glass matrix elements can be approximated by summing the Si and B in solution in mM/L.[21,22]

The leachate chemistry can, therefore, be modeled as a mixture of weak acids in equilibrium with their alkali metal salts or strong bases. The [SB] minus [WA] equilibria is calculated by the difference in the [SB] and the [WA]. The excess strong base, $[SB]_{ex}$, is [SB]-[WA]>0. Where

$$\underbrace{[SB]-[WA]}_{leachate} \equiv \underbrace{[Na + Li + K]_{mM}}_{leachate} - \underbrace{[B + Si]_{mM}}_{leachate} \qquad (3)$$

The leachate concentrations, when expressed in terms of [SB]-[WA] versus measured pH of PCT-A and PCT-B leachates are not linear. Where there is little change in the measured pH, e.g. Figure 5a or Figure 5b, there are wide changes in corresponding [SB]-[WA] equilibria. When [SB]-[WA] buffer capacity of a leachate solution is exceeded (Figure 5b, d, f), e.g., for glasses of poorer durability and/or higher alkali content which creates $[SB]_{ex}$, glasses return to an accelerated rate of leaching. For durable glasses, the leachates are buffered at a neutral [SB]-[WA] of ~0. The corresponding sharp break in the pH versus [SB]-[WA] curve at pH of ~10.8 at ambient temperature corresponds to the pH in the vicinity of the defense HLW high iron (Purex) glasses. Glasses such as the HLW standard Environmental Assessment (EA) glass are in the pH range where excess base, $[SB]_{ex}$, dominates. This is in agreement with the pH of 10.7 at ambient temperature which is the value noted by Muller, et. al. [16] and Ribet et.al.[17] that distinguishes between glasses that return to an accelerated dissolution rate and those that continue leaching at a steady state rate. Due to the non-linearity of the pH with the [SB]-[WA] equilibria, the $[SB]_{ex}$ is the key parameter responsible for the hydrogel transformation to zeolite and not pH.

Six glasses are given as examples of the impact of the [SB]-[WA] leachate equilibria in Figure 5. The SON68 glass (Figure 5a) continues to generate [WA] with continuing leaching test duration in PCT-A and PCT-B. The Waste Compliance Plan (WCP) HM glass, which is a high Al glass, continues to be buffered with continuing leaching test duration (Figure 5c) as does the WCP Batch 1 glass (Figure 5e). Conversely, the AFCI glass generates excess [SB] at 7 day test duration and continues to generate [SB]ex in the leachate with the test duration (Figure 5b). The EA glass (Figure 5d) and WCP Purex glass (Figure 5f) perform similarly to the AFCI glass and all three of these glasses return to an accelerated rate of leaching with test duration (Figure 4). The buffered leachates for SON68, HM and Batch 1 glass do not return to an accelerated rate of leaching (Figure 4).

179

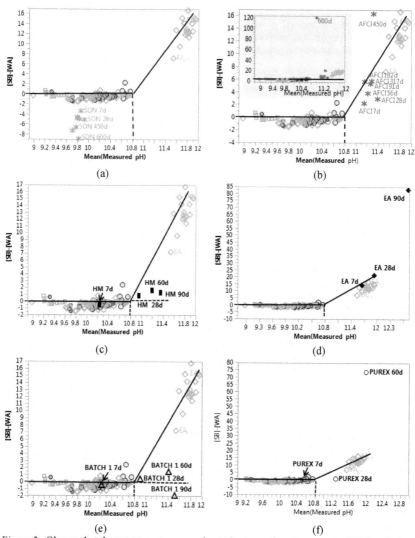

Figure 5. Glasses that do not return to an accelerated rate continue to generate WA in solution with time (a) or continue to be buffered at [SB]-[WA] ~0 (c and e). Glasses that do return to an accelerated rate continue to generate SB in solution with time (b, d, and f). Gray shaded data is 7-day PCT-A data from reference 21. Outlined circles are 7 day PCT-A data for WCP Purex glass that defines the break in the leachate buffer curve from reference 21.

Figure 6 shows a composite of the durability data for all seven HLW glasses added to ALTGLASS and discussed in this study. All of the glasses that return to an accelerated rate, i.e. EA, WCP Purex and AFCI glass, have [SB]$_{ex}$, i.e. [SB]>>0 as time of corrosion testing continues. All of the glasses that either produce additional [WA] as corrosion testing continues or continue to be buffered, i.e. WCP HM, Blend 1, Batch 1 and SON 68, do not return to an accelerated rate of dissolution with time.

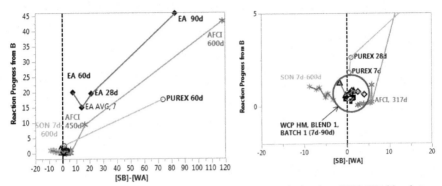

Figure 6. Reaction Progress for the seven HLW glasses versus the leachate [SB]-[WA] leachate equilibria shown at full scale (left) and the detail shown as an inset (right).

Alkali in the Hydrogel

The glass corrosion front parallels the original surface of the glass. If a glass leaches congruently, all the elements would leach into the leachate at the same rate as B. However, many elements get sequestered in the hydrogel leached layer and never make it into the leachate. Therefore, by the law of mass action, one can calculate the composition (in mg/L) of element "i" in the gel layer which includes the newly formed gel and the gel being transformed in situ into secondary phases and then convert to millimoles/L (mM/L) of element "i" in the gel layer for modeling.

$$C_{i(gel)} = \left(\left(\frac{(\text{Re}\,action\,Pr\,ogress)_B}{100} \right) * \left(C_{i(glass)} * 1000 \right) \right) - C_{i(leachate)} \qquad (4)$$

$$\underbrace{\qquad\qquad\qquad\qquad\qquad\qquad\qquad\qquad}_{Glass\,component\,that\,could\,have\,dissolved\,in\,the\,solution} \quad \underbrace{\qquad\qquad}_{Glass\,component\,found\,in\,solution}$$

where

Reaction Progress	=	% of glass leached
Ci (gel)	=	concentration of element "i" in the gel that could have gone into the leachate at this reaction progress, mg$_i$/L
C$_i$(leachate)	=	concentration of element "i" in the leachate, mg$_i$/L,
Ci(glass)*1000	=	concentration of element "i" in the glass that should have dissolved into the leachate at this reaction progress, mg$_i$/L

Gel compositions were calculated for the seven HLW glasses examined in this study. As found with the previous 12 HLW glasses assessed [11,12] silica is the major component of the gel layer, followed by $\Sigma alkali_{gel} = Na_{gel} + Li_{gel} + K_{gel}$ for glasses that resume leaching at an accelerated rate, i.e. the EA and Purex glasses as the AFCI glass had no alkali in the gel layer. For glasses that resume leaching at an accelerated rate the Li_{gel} dominates over Na_{gel} and K_{gel}. For the glasses that did not resume leaching at an accelerated rate, i.e. Batch 1, Blend 1, and HM, Fe_{gel} dominated followed by Al_{gel} and there was little Si_{gel} or $\Sigma alkali_{gel}$ in the leached layer after 28 days of testing.

Interactions of [SB]-[WA] Equilibria and Hydrogel

Excess alkali or poorly bound alkali in the parent glass must be the source of (1) the buildup of $\Sigma alkali_{gel}$ in the accelerated resumption glasses and (2) the buildup of $[SB]_{ex}$ in the leachates of the accelerated resumption glasses. The interaction of the $[SB]_{ex}$ in the leachate interacts with the silica-rich aluminosilicate hydrogel layer created during leaching and stabilizes zeolites rather than clay mineral assemblages. The mechanism for zeolite formation in PCT-A and PCT-B, therefore, is the same as that used in the industrial synthesis of zeolites from aluminosilicate gels in the presence of excess (100-300%) NaOH.[23] The operative mechanism for the zeolite formation therefore depends on the presence of excess strong base, $[SB]_{ex}$, in the leachates, whereas, clay formation occurred in glasses where the leachates were buffered.

Based on previous analytic analyses of the leach layer composition [11] of the "return" vs. "no return" grandchildren populations in the ALTGLASS database, the following preliminary rate determining reactions were be hypothesized. These reactions are supported by the current findings.

For Fe-Si rich hydrogels that do not return to the accelerated rate:

$$\underbrace{2Na_2O \bullet 2Fe_2O_3 \bullet 8SiO_2}_{hydrogel} + 15H_2O \rightarrow \underbrace{2(Fe^{+3})_2 Si_2O_5(OH)_4 \bullet 2H_2O}_{hisingeriteprecursorgel} + \underbrace{4NaOH + 4H_4SiO_4}_{bufferedleachate} \quad (5)$$

Note that in Equation 5 that equal moles of [SB] and [WA] are generated on the product side of the reactions which keeps the leachates buffered which is what is observed.

For Al-Si rich hydorgels that do return to the accelerated rate and there is alkali in the hydrogel and excess alkali is present as $[SB]_{ex}$ in the leachates at 100-300% excess (Na,Li,K)OH:

$$\underbrace{1 Na_2O \bullet Al_2O_3 \bullet 4SiO_2 \bullet 2H_2O}_{hydrogel} + \underbrace{2LiOH}_{leachateexcessbase} \rightarrow \underbrace{(Na,Li)Al_2Si_4O_{12} \bullet 2H_2O}_{analcimeprecursorgel} + \underbrace{2(Na,Li)OH}_{leachate} \quad (6)$$

If the hydrogel is polyamorphous, which may be why clay minerals sometimes form first on glass leach layers followed by zeolitic minerals, the hydrogel structure may have a [SB]-[WA] dependency, i.e. in buffered leachates the following reaction likely occurs

$$3Na_2O \bullet \underbrace{2Fe_2O_3 \bullet 2Al_2O_3 \bullet 12SiO_2}_{hydrogel} + 17H_2O \rightarrow$$

$$\underbrace{2(Fe^{+3})_2 Si_2O_5(OH)_4 \bullet 2H_2O}_{his\,in\,gerite} + \underbrace{2(Al_2O_3 \bullet SiO_2)}_{hydrogel} + \underbrace{6NaOH + 6H_2SiO_3}_{bufferedleachate} \qquad (7)$$

Equation 7 produces an aluminosilicate hydrogel which may cause an analcime layer to form on top of or subsequent to a clay layer as observed experimentally.[24]

In a [SB]$_{ex}$ or excess NaOH dominated leachate the following reaction likely occurs

$$3Na_2O \bullet \underbrace{2Fe_2O_3 \bullet 2Al_2O_3 \bullet 12SiO_2}_{hydrogel} + 2LiOH + 7H_2O \rightarrow$$

$$\underbrace{2NaLiAl_2Si_4O_{12} \bullet 2H_2O}_{analcimeprecursorgel} + \underbrace{2(Fe_2O_3 \bullet SiO_2)}_{hydrogel} + \underbrace{4NaOH + 2H_2SiO_3}_{SBdomin\,atedleachate} \qquad (8)$$

CONCLUSIONS

The data presented in this study is confirmatory of previous work on the ALTGLASS database [12] where it was shown that high alkali glasses created excess alkali in the gel layer and in the leachate and that the high alkali was primarily responsible for the return to the accelerated rate. Li$_{gel}$ was found to cause a stronger impact than the other alkalis.

The excess alkali in the leachates of glasses that resume leaching at an accelerated rate interact with the hydrogel layer created during leaching and stabilize zeolites rather than clay mineral assemblages. The mechanism for zeolite formation, therefore, is the same as that used in the industrial synthesis of zeolites from aluminosilicate gels in the presence of excess NaOH.[23] The operative mechanism for the zeolite formation therefore depends on the presence of excess stong base, [SB]$_{ex}$, in the leachates, whereas, clay formation occurs in buffered leachates or leachates with [WA]$_{ex}$. These data indicate that the gel layer transforms to secondary phases in situ in response to interactions with the chemistry of a continuously evolving leachate.

The impact of the [SB]-[WA] leachate equilibria in closed system durability tests in deionized water raises the question as to the impact of silicate rich, i.e. [WA] rich, groundwaters on the return to the accelerated rate of dissolution for high alkali containing glasses. Comparative testing is needed for the same reference glasses in silicate rich groundwater such as the J13 Yucca Mtn. groundwater or granitic or basaltic repository groundwater.

ACKNOWLEDGMENTS

This work is part of a joint EM-NE-SC-International Technical Evaluation of Alteration Mechanism (I-TEAM) Glass Corrosion Program to develop the data and understanding necessary for an international consensus on the behavior of glass over geologic time scales in a variety of disposal environments.[2] This work was supported by the U.S. Department of Energy, Office of Environmental Management, EM-31, under Contract DE-AC09-08SR22470 and the U.S. Department of Energy Office of Nuclear Energy, under Contract DE-AC02-06CH11357.

REFERENCES

1. P. Van Iseghem and B. Grambow, Sci. Basis for Nuclear Waste Management XI, J.J. Apted and R.E. Westerman (Eds.), Mat. Res. Soc., Pittsburgh, PA, 631 (1987).
2. Ryan, J.V., Ebert, W.L., Icenhower, J.P., Schreiber, D.K., Strachan, D.M., and Vienna, J.D., WM12 Paper #12303 (2012).
3. P. VanIseghem, M. Aertsens, S. Gin, D. Deneele, B. Grambow, P. McGrail, D. Strachan, and G. Wicks, EUR23097 (2007).
4. A.C. Lasaga, In *Chemical Weathering Rates of Silicate Minerals* (White, A. F., and Brantley, S., Eds.), pp 353-406, Mineralogical Society of America, Washington, DC (1995).
5. B. Grambow and R. Muller, J. of Nuclear Materials 298, 112-124 (2001).
6. P. Frugier, S. Gin, Y. Minet, T. Chave, B. Bonin, N. Godon, J.-E. Lartigue, P. Jollivet, A. Ayral, L. DeWindt, and G. Santarini, Jour. Nucl. Materials, 380, 8-21 (2008).
7. P. Frugier, T. Chave, S. Gin, and J.-E.Lartigue, Jour. Nucl. Materials, 392, 552-567 (2009).
8. S.V. Raman, Physics and Chem. of Glasses, 42[1], 27-41 (2001).
9. T.A. Abrajano, J.K. Bates, A.B. Woodland, J.P. Bradley, and W.L. Bourcier, Clays and Clay Minerals, 38[5], 537-548 (1990).
10. E.H. Oelkers, Geochim. Cosmochim. Acta, 65 [21], 3703-3719 (2001).
11. C.M. Jantzen, C.L. Crawford, J.M. Pareizs, and J.B. Pickett, "Accelerated Leach Testing of GLASS (ALTGLASS): Part I. Waste Glass Hydrogel Compositions and Accelerated Dissolution," accepted Intl. J. App. Glass Sci. (2015).
12. C.M. Jantzen, C.L. Crawford, J.M. Pareizs, and J.B. Pickett, Jantzen,C.M., Crawford, C.L, Pareizs, J.M, and Pickett, J.B., "Accelerated Leach Testing of GLASS: II. Leachate-Hydrogel Interactions, Glass Structure, and Accelerated Dissolution" accepted Intl. J. App. Glass Sci. (2015).
13. C.M. Jantzen, N.E. Bibler, D.C. Beam, and M.A. Pickett, U.S. DOE Report WSRC-TR-92-346, Rev.1, 92p (1993). www.osti.gov/scitech/search
14. S.L. Marra, A.L. Applewhite-Ramsey, and C.M. Jantzen, Ceramic Trans. V. 23, Am.Ceram. Soc., Westerville, OH, 465 (1991).
15. C.M. Jantzen and C.L. Crawford, Accelerated Leach Testing of GLASS (ALTGLASS)-Version 2.1 Database, SRNL-L3100-2014-00229 and FCRD-SWF-2014-000249, 111 pp. (2014).available electronically from the authors
16. I. S. Muller, S. Ribet, I.L. Pegg, S. Gin and P. Frugier, Ceram. Trans., V.176, Am.Ceram. Soc., Westerville, OH, 191 (2006).
17. S. Ribet, I.S. Muller, I.L. Pegg, S. Gin, and P. Frugier, Sci. Basis for Nucl. Waste Mgt., XXVIII, Mat. Res. Soc., Warrendale, PA, 309 (2004).
18. D.M. Strachan, U.S. DOE Report PNNL-12074 (1998). www.osti.gov/scitech/search
19. ASTM C 1285 – 14, "Standard Test Methods for Determining Chemical Durability of Nuclear, Hazardous, and Mixed Waste Glasses and Multiphase Glass Ceramics: The Product Consistency Test (PCT)".
20. E.M. Pierce, L.R. Reed, W.J. Shaw, B.P. McGrail, J.P. Icenhower, C.F. Windisch, E.A. Cordova, and J. Broady, Geochim. Comochim. Acta, 74, 2634 (2010).
21. C.M. Jantzen, J.B. Pickett, K.G. Brown, T.B. Edwards, and D.C. Beam,U.S. Patent 5,846,278 (1998).
22. C.M. Jantzen, K.G. Brown, and J.B. Pickett, Intl. J. App. Glass Sci., 1 [1], 38-62 (2010).
23. R.M. Barrer, "Hydrothermal Chemistry of Zeolites," Academic Press, New York,360pp. (1982).
24. T.A. Abrajano, J.K. Bates, A,B. Woodland, J.P. Bradley, and W.L. Bourcier, Clays and Clay Miner. 38, 537-548 (1990).

Mater. Res. Soc. Symp. Proc. Vol. 1744 © 2015 Materials Research Society
DOI: 10.1557/opl.2015.334

Solution Composition Effects on the Dissolution of a CeO_2 analogue for UO_2 and ThO_2 nuclear fuels

Claire L. Corkhill[1], Martin C. Stennett[1] and Neil C. Hyatt[1]

[1]Immobilisation Science Laboratory, Department of Materials Science and Engineering, University of Sheffield, S1 3JD, U.K.

ABSTRACT

This study investigates the dissolution of CeO_2, an isostructural analogue for UO_2 and ThO_2, which was synthesized to closely approximate the microstructure of a spent nuclear fuel matrix. Dissolution of CeO_2 particles was performed in simplified solutions representative of saline, near-neutral and alkaline ground waters that may be encountered in geological disposal scenarios, and in acidic medium for comparison. The normalized mass loss of cerium was found to be significantly influenced by the formation of colloidal particles, especially in the near-neutral and alkaline solutions investigated. The normalized dissolution rate, $RL_{(Ce), k}$ (g m^{-2} d^{-1}), in these two solutions was found to be similar, but significantly lower than in a nitric acid medium. The activation energies based on the normalized release rate of cerium, at 40°C, 70°C and 90°C in each solution, were in the range of 24 ± 3 kJ mol^{-1} to 27 ± 7 kJ mol^{-1}, indicative of a surface-mediated dissolution mechanism. The mechanism of dissolution was postulated to be similar in each of the solutions investigated, and further work is proposed to investigate the role of carbonate on the CeO_2 dissolution mechanism.

INTRODUCTION

In the safety case for the geological disposal of spent nuclear fuel, which is composed primarily of UO_2, the release of radioactivity to the geosphere will be controlled by the dissolution of the fuel in ground water. Investigations of the dissolution behaviour of UO_2 can be hindered by the rapid oxidation of U(IV) to U(VI), therefore alternative, non-redox active analogues may assist the determination of dissolution rates.

Cerium dioxide (CeO_2) is isostructural to UO_2, sharing the same fluorite structure (space group Fm-3m). It is also isostructural to ThO_2, which has received much recent interest as a potential future fuel [1]. Ce(IV) is the highest oxidation state achievable for CeO_2, therefore oxidation is not a concern during dissolution experiments. Recent work has been performed to investigate the dissolution of CeO_2, synthesised to closely approximate the mineral structure characteristics of fuel grade UO_2 (e.g. grain size, crystallographic orientation and sintered density) [2, 3]. These investigations were performed in acidic media, which are useful for the elucidation of reaction kinetics on laboratory timescales, but are not representative of geological disposal groundwater compositions. Because solution composition has been shown to greatly affect UO_2 and ThO_2 dissolution rates [4], the focus of this investigation is the dissolution of CeO_2 in simplified solutions representative of groundwater previously used in studies of spent nuclear fuel dissolution [5]. By understanding the effects of solution composition on the dissolution rate, we aim to evaluate the suitability of CeO_2 as an analogue for UO_2 and ThO_2 nuclear fuels.

METHODOLOGY

CeO_2 pellets, prepared according to [2], were crushed and washed according to the Product Consistency Test ASTM standard [6] to a 25 – 50 μm size fraction. Prior to use, the particles were inspected by Scanning Electron Microscopy to ensure no fine particles < 1 μm in size remained. Surface area was determined using the BET method, with a nitrogen adsorbate, the resulting surface area was found to be 0.304 ± 0.001 m^2 g^{-1}. Samples of 0.1 g of CeO_2 were placed in 50 mL PTFE vessels and filled with 40 mL of alteration solution. Solutions, prepared using ultra-high quality water (18 MΩ cm^{-1}) were: 0.01 M HNO_3; a simplified saline, carbonate-containing solution of NaCl (0.01 M) and $NaHCO_3$ (0.002 M) and; a high ionic strength 0.1 M NaOH solution. The pH of these solutions at room temperature [pH (23°C)] were pH (23°C) = 2.0 ± 0.2, pH (23°C) = 8.5 ± 0.2 and pH (23°C) = 11.8 ± 0.2, respectively. Dissolution experiments were performed in triplicate at 40 °C, 70 °C and 90 °C, with duplicate blanks. Sampling was conducted at regular intervals from 0 to 35 days. An aliquot (1.2 mL) of each sample was removed and filtered (0.22 μm) prior to analysis by ICP-MS (Agilent 4500 Spectrometer). Leaching is expressed as the normalised elemental leaching $NL_{(Ce)}$ (g m^{-2}) according to:

$$NL_{(Ce)} = {m_{(Ce)}}/{SA/V} \qquad (1)$$

where m_{Ce} is the total amount of cerium released in the solution at each sampling time and SA/V is the surface area to volume ratio. The normalised element leaching rate $RL_{(Ce)}$ (g m^{-2} d^{-1}) was determined by:

$$RL_{(Ce)} = NL_{(Ce)} \cdot \Delta t \qquad (2)$$

where Δt is the leaching time in days.

RESULTS AND DISCUSSION

Effect of filtration on dissolution rate

The dissolution of CeO_2 can be described by two distinct dissolution regimes, as shown in Figure 1a for CeO_2 in nitric acid. The first, rapid regime is far from solution saturation, thus represents the kinetically-controlled dissolution rate, $RL_{(Ce),k}$. The second regime is representative of near-solution saturation. This is much less rapid due to thermodynamic effects that occur close to equilibrium, and is denoted $RL_{(Ce),t}$. Prior to investigation of the effects of solution composition on dissolution rate, the effect of solution filtration was determined.

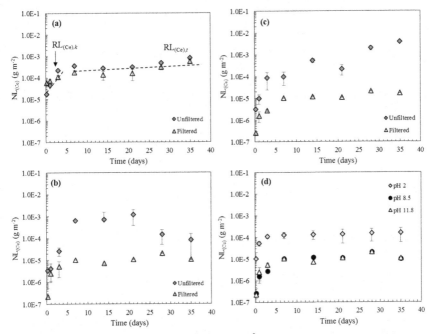

Figure 1. Normalized mass loss of cerium ($NL_{(Ce)}$, g m^{-2}) resulting from the dissolution of CeO_2 particles, at 90°C, in **(a)** 0.01 M HNO_3 [pH(23°C) = 2]; **(b)** 0.01 M NaCl / 0.002 M $NaHCO_3$ [pH(23°C) = 8.5]; and **(c)** 0.01 M NaOH [pH(23°C) = 11.8] as a function of filtration, and **(d)** the comparative normalized mass loss of cerium from each filtered solution.

Normalized cerium concentrations in each of the sample solutions, at 90°C, as a function of time and filtration are shown in Figure 1 and Table 1. Filtration of solutions arising from CeO_2 dissolution in nitric acid gave a kinetic dissolution rate ($RL_{(Ce),k}$) that was 20 times lower than the unfiltered samples in the same solution (Fig. 1a). The cerium concentrations in filtered NaCl / $NaHCO_3$ and NaOH solutions were up to 60 times lower than those observed in unfiltered solutions (Figs. 1b and c). This indicates that colloidal cerium species were formed during the dissolution of CeO_2, particularly in near-neutral to alkaline pH solutions.

Analysis of the resulting solution using geochemical modelling (PHREEQC) indicated that a range of aqueous Ce(IV) species, including $Ce(OH)_2^{2+}$, $CeOH^{3+}$ and $Ce(CO_3)^{2-}$ may be formed in the NaCl / $NaHCO_3$ and NaOH solutions. It is well known that $Th(IV)O_2$ dissolution in synthetic groundwater media also generates a wide variety of hydroxide and carbonate colloid species, through a process of hydrolysis and polynucleation [7]. This process differs from that of

UO$_2$, which also forms complexed carbonate species, but this is only effective if U(VI) is present at the surface [5].

Table 1. Normalized dissolution rates of Ce in the kinetic (RL$_{(Ce),k}$) and near-solution saturation (RL$_{(Ce),l}$) regimes, as a function of solution composition and filtration.

Solution Composition	RL$_{(Ce),k}$ (g m^{-2} d^{-1})	RL$_{(Ce),l}$ (g m^{-2} d^{-1})
Unfiltered		
0.01 M HNO$_3$	$(3.49 \pm 0.2) \times 10^{-4}$	$(5.00 \pm 0.8) \times 10^{-6}$
0.01 M NaCl / 0.002 M NaHCO$_3$	$(1.85 \pm 0.1) \times 10^{-5}$	$(1.16 \pm 0.1) \times 10^{-4}$
0.01 M NaOH	$(9.40 \pm 1.2) \times 10^{-5}$	$(7.84 \pm 2.1) \times 10^{-7}$
Filtered		
0.01 M HNO$_3$	$(1.76 \pm 0.4) \times 10^{-5}$	$(1.11 \pm 0.8) \times 10^{-6}$
0.01 M NaCl / 0.002 M NaHCO$_3$	$(1.47 \pm 0.1) \times 10^{-6}$	$(1.32 \pm 0.1) \times 10^{-6}$
0.01 M NaOH	$(1.50 \pm 0.1) \times 10^{-6}$	$(1.70 \pm 0.1) \times 10^{-6}$

The normalized cerium concentrations from each of the filtered solution compositions, as a function of time, are shown in Figure 1d. The dissolution rates for CeO$_2$ in acid were at least one order of magnitude greater than those for the near-neutral to high pH solutions, giving an RL$_{(Ce),k}$ of $(1.76 \pm 0.40) \times 10^{-5}$ g m^{-2} d^{-1} (Table 1). The dissolution rates in NaCl / NaHCO$_3$ and NaOH solutions were very similar, with RL$_{(Ce),k}$ values of $(1.47 \pm 0.01) \times 10^{-6}$ g m^{-2} d^{-1} and $(1.50 \pm 0.04) \times 10^{-6}$ g m^{-2} d^{-1}, respectively. This suggests that, unlike ThO$_2$, complexation of cerium by carbonates did not enhance the rate of dissolution of CeO$_2$, however, further work using a range of carbonate concentrations is required to confirm this.

The effect of temperature on dissolution rate

The influence of temperature on the dissolution rate of CeO$_2$ in each of the solution compositions was determined at 40°C, 70°C and 90°C. The observed dependence of temperature on the dissolution rate can be described by the Arrhenius equation:

$$r = A^{\frac{-E_a}{RT}} \tag{3}$$

where r is the dissolution rate (g m^{-2} d^{-1}), A is the Arrhenius parameter (g m^{-2} d^{-1}), E$_a$ is the activation energy (kJ mol^{-1}), R is the ideal gas constant in J (mol K^{-1}) and T is the temperature (K). Using the Arrhenius dependence of dissolution rate on temperature, linear regression of the dissolution data in the kinetic dissolution regime (RL$_{(Ce),k}$) as a function of 1/T was used to calculate the activation energy for CeO$_2$ in each of the solution compositions. The resulting values, with 2σ errors and the correlation coefficient of the linear regression are shown in Table 2.

Table 2. Calculated activation energies (E_a, kJ mol^{-1}) for CeO_2, and the correlation coefficient (r^2) derived from regression analysis based on the normalized release of cerium ($RL_{(Ce),k}$) in each of the solution compositions investigated.

Solution Composition	E_a (kJ mol^{-1})	r^2
0.01 M HNO$_3$	24.4 ± 3	0.99
0.01 M NaCl / 0.002 M NaHCO$_3$	26.0 ± 5	0.96
0.01 M NaOH	27.4 ± 7	0.99

The dependence of normalized dissolution rate on temperature gives activation energies in the range of 24 ± 3 kJ mol^{-1} to 27 ± 7 kJ mol^{-1}. These values are in the range proposed by Lasaga [8] to be consistent with a surface-controlled dissolution mechanism. This is in agreement with previous studies of CeO_2 and other fluorite-type dioxides (e.g. ThO_2 and PuO_2) during dissolution, which reported activation energies in the range of 20 to 37 kJ mol^{-1} [9, 10]. The activation energies for CeO_2 derived from each solution were extremely similar, indicating that the dissolution mechanism of CeO_2 was the same, regardless of solution composition. However, the dissolution rates in acidic media were clearly greater than in near-neutral to alkaline solutions. The activation energy arising from CeO_2 dissolution in the carbonate-containing solution was similar to the other solutions without carbonate, which provides further evidence that complexation of cerium by carbonate species does not appear to influence the dissolution mechanism in these experiments. However, it should be noted that these experiments were performed in equilibrium with the air, therefore further experiments under conditions of CO_2-exclusion are necessary to confirm the role of carbonate in CeO_2 dissolution.

CONCLUSIONS

We have investigated the dissolution of a CeO_2 analogue for UO_2 and ThO_2 fuels in solution compositions representative of simplified ground water solutions. The dissolution rate of CeO_2 was found to be strongly influenced by colloidal species, especially in near-neutral to alkaline pH solutions. The CeO_2 dissolution rate in a saline, bicarbonate-buffered solution was found to be extremely similar to that in a strongly alkaline solution, but dissolution rates were an order of magnitude greater in acidic solution. The activation energy of CeO_2 in each solution was found to be similar, and in the range suggestive of a surface-controlled dissolution mechanism. These results suggest that, unlike $Th(IV)O_2$, $Ce(IV)O_2$ dissolution was not strongly influenced by carbonate complexation, however further work investigating the dissolution of CeO_2 in a range of carbonate concentrations, under CO_2-exclusion is necessary to confirm this observation.

ACKNOWLEDGMENTS

The research leading to these results has received funding from the European Atomic Energy Community's Seventh Framework Programme (FP7) under grant agreement No. 269903, The

REDUPP (REDucing Uncertainty in Performance Prediction) project. We are grateful to Dr. Virginia Oversby and Dr. Lena Z. Evins for invaluable discussion and support throughout the project. CLC is grateful to The University of Sheffield for the award of a Vice Chancellor's Fellowship and NCH acknowledges support from the Royal Academy of Engineering and the Nuclear Decommissioning Authority. We also wish to thank EPSRC for funding under grant reference EP/L014041/1 (DISTINCTIVE project) and also EPSRC and The University of Sheffield Knowledge Transfer Account, under grant reference EP/H500170/1.

REFERENCES

1. V. Dekoussar, G. R. Dyck, A. Glperin, C. Ganguly, M. Todosow and M. Yamawaki. IAEA-TECDOC-1450 (2005).
2. M. C. Stennett, C. L. Corkhill, L. A. Marshall and N. C. Hyatt. *J. Nucl. Mater.* **432**, 182 (2013).
3. C. L. Corkhill, E. Myllykylä, D. J. Bailey, S. M. Thornber, J. Qi, P. Maldonado, M. C. Stennett, A. Hamilton and N. C. Hyatt. *Appl. Mater. Interfac.* **6**, 12279 (2014).
4. M. Altmaier, V. Neck, R. Muller and Th. Fanghanel. *Radiochim. Acta* **93**, 83 (2005).
5. V. M. Oversby. *SKB Tech. Rep.* **TR-99-22** (1999).
6. American Society for Testing and Materials, Philadelphia, PA. ASTM C 1285-02 (2008).
7. D. Rai, D. A. Moore, C. S. Oakes and M. Yui. *Radiochim. Acta* **88**, 297 (2000)
8. A. C. Lasaga, *in Kinetics of Geochemical Processes* edited by A. C. Lasaga and P. H. Ribbe (Reviews in Mineralogy, Washington DC, 1981) p.1
9. L. Claparede, N. Clavier, N. Dacheux, A. Mesbah, J. Martinez, S. Szenknect and P. Moisy, *Inorg. Chem.* **50**, 11702 (2011).
10. G. Heisbourg, S. Hubert, N. Dacheux and J. Purans. *J. Nucl. Mater.* **335**, 5 (2004).

Storage and Disposal of Nuclear Waste

Mater. Res. Soc. Symp. Proc. Vol. 1744 © 2015 Materials Research Society
DOI: 10.1557/opl.2015.313

Deep Borehole Disposal Research: What have we learned from numerical modeling and what can we learn?

Karl P Travis and Fergus G F Gibb
Immobilisation Science Laboratory, Department of Materials Science & Engineering, The University of Sheffield, Sheffield S1 3JD, United Kingdom.

ABSTRACT

Geological disposal of HLW and spent nuclear fuel (SNF) in very deep boreholes is a concept whose time has come. The alternative – disposal in a mined, engineered repository is beset with difficulties not least of which are the constraints placed upon the engineered barriers by the high thermal loading. The deep borehole concept offers a potentially safer, faster and more cost-effective solution. Despite this, international interest has been slow to materialize, largely due to perceived problems with retrievability and uncertainty about the ability to drill accurate vertical holes with diameters greater than 0.5 m to a depth of 4-5 km. The closure of Yucca Mountain and the subsequent recommendations of the Blue Ribbon Commission have lead to a renewed interest in deep borehole disposal (DBD) and the US DoE has commissioned Sandia National Labs, working with industrial and academic partners (including the University of Sheffield), to undertake a program of R&D leading to a demonstration borehole being drilled somewhere in the continental USA by 2016.
    In this paper, we focus on some of the key safety and engineering features of DBD including methods of sealing the boreholes, sealing and support matrices for the waste packages. Numerical modeling has, and continues to play, a significant role in expanding and validating the DBD concept. We report on progress in the use of modeling in the above contexts, paying particular attention to constraints on the engineering materials resulting from high heat loading.

INTRODUCTION

The deep borehole disposal (DBD) concept involves the drilling of a vertical hole to a depth of 4-5 km into the granitic basement of the continental crust. The hole is then lined with steel casing and the lower 1-2 km filled with waste packages together with a sealing and support matrix (e.g. a cementitious grout). Throughout the disposal zone (DZ) the casing is perforated for various reasons including weight reduction. The borehole is then backfilled and sealed above the disposal zone. Hole diameters can vary from 8.5 to 24 inches, but this is largely dependent on current drilling envelopes and the type of waste being considered[1].
    Although DBD is a true multi-barrier concept, its chief advantage over a repository is an order of magnitude greater geological barrier. At depths of 4-5 km, lateral movement of groundwater through igneous rock such as granite is generally limited due to the very low hydraulic conductivities usually found at such depths while density stratification has frequently resulted in highly saline groundwaters being isolated from the near surface groundwaters for millennia. Apart from safety, speed of implementation is another major advantage offered by DBD over a repository; a single borehole could be drilled, cased, filled and sealed in a little under 3 years. In the UK, a proposal to dispose of HLW in the same repository as that used for ILW would not see the first HLW containers being emplaced until at least the year 2075. A

report by Radioactive Waste Management (RWM) Ltd suggests that this date could be brought forward to the year 2040 if instead, the UK's HLW was to be disposed of in deep boreholes [2]. The recent report by the International Panel for Climate Control (IPCC) suggests that the electricity generated from nuclear, renewables and fossil fuels with carbon capture and storage (CCS) technologies would need to increase from its current share of around 30% of the total electricity supply to 80% by 2050 if the target of zero fossil fuel use is to be achieved. Increasing the nuclear fission baseload would entail a much greater quantity of spent fuel to be ultimately disposed of, placing an even greater thermal stress on a repository and vastly increasing its footprint.

DBD offers a significant cost advantage over a repository concept for disposal of spent fuel. At the time of the closure of the Yucca Mountain project in 2010, the US government had spent an estimated $15 billion. Boreholes on the other hand are estimated to cost $60M for drilling the first hole [1], but less for subsequent holes. Any cost comparison for disposal in a borehole versus disposal in a GDF such as the Swedish KBS-3 concept should be made on a per tonne of heavy metal basis. Taking LWR spent fuel, such a comparison predicts that this cost would be $115K/tHM for disposal in a borehole compared with $550K/tHM in a KBS-3 GDF. That is, disposal via boreholes would be around five times cheaper.

BOREHOLE DESIGNS

A key advantage of DBD is its versatility; the basic design concept can be easily modified to enable the disposal of a wide range of different waste packages. Three main variations of the concept have been proposed by workers at the University of Sheffield. Two low temperature schemes: LTVDD 1, LTVDD 2, and a high temperature variant: HTVDD. Figure 1 is a schematic diagram showing the principle differences between these variants.

The key difference between LTVDD 1 and LTVDD 2 lies in the sealing and support matrix (SSM). The SSM's primary function is to act as a seal, preventing ingress of groundwater, which might corrode the waste containers, and closing off an easy route for any escaped radionuclides back to the surface through the annulus. The secondary function is to provide mechanical support for each waste package against the compressive load of the overlying packages. LTVVD 1 is designed for wasteforms which have a relatively low heat output and thus a cementitious grout is a suitable choice for the SSM. Cement is widely used in the oil and gas industry to seal boreholes mainly because of its low cost and relative ease of deployment. However, DBD brings with it new scientific and engineering challenges for a cement-based SSM: the setting time of any grout must be delayed long enough to enable it to flow and fill all available space in the DZ prior to the next package being delivered down-hole. This must be achieved despite elevated temperatures of upwards of $100°C$ and hydrostatic pressures (while the borehole remains open) of around 50 MPa. The particles in the grout must be prevented from dispersing too much in the borehole fluid. Adding the necessary retarding agents, plasticizers *etc.* to yield a grout with these properties and at the same time having a consistency which is below the limit of pumpability is a difficult task. We are currently engaged in a program of research to develop a range of such grouts for use in LTVDD 1.

194

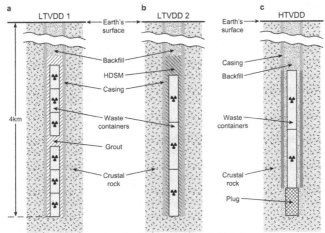

Figure 1. Schematic diagram showing the three main Sheffield design variants of DBD: (a) LTVDD 1, (b) LTVDD 2, and (c) HTVDD.

Several years ago, we proposed a novel SSM which comprised a lead-tin alloy, deployed as a fine shot. We termed this new SSM: High Density Support Matrix (HDSM) [3]. The basic idea behind the HDSM is to utilize the radiogenic heat provided by the waste packages to melt the alloy, providing a dense fluid able to perfectly fill the voids in the DZ, which upon eventual cooling, will 'solder' the waste packages into the hole. The HDSM is particularly useful for disposal of SNF which gives rise to very dense waste packages. The density of the alloy is such that the packages are slightly negatively buoyant and therefore would sink slowly under their own weight through the molten alloy. LTVDD 2 was designed to dispose of SNF, both as whole assemblies and as consolidated fuel pins.

The best SSM is granite itself; the HTVDD concept is based on this idea. Waste packages containing very young, hot SNF (for example) with a high heat output, could generate sufficient heat to melt the backfill (crushed rock) and solid granite in the DZ. The molten granite will flow into any voids, sealing the lower reaches of the hole. Once the down-hole temperature drops below the liquidus for granite at the ambient pressure, the granite will begin to recrystallize, forming a hard rock seal. To prevent the first waste package melting through the bottom of the hole, a refractory plug is emplaced prior to beginning waste deployment.

HEAT FLOW MODELING

Mathematical models play a crucial role in borehole design. Mechanical models enable one to: determine the effect of lithostatic stresses on the waste packages, the integrity of the rock in the engineering disturbed zone and any mechanical sealing devices employed, as well as the flow of fluids through the borehole annulus, and voids in the rock. Thermal models on the other hand, allow one to determine: the optimum waste inventory for a given hole, the most appropriate SSM to deploy, the maximum temperatures that will be attained in and around the waste disposal zone and the time taken to reach them. Building a thermo-mechanical-hydraulic model capable of dealing with all of the above and able to account for the geochemistry is a desirable end goal. However, such a model is extremely complicated and rapid progress is more easily attained by using more limited models which, for instance, treat the flow of heat by considering conduction only – convective transport is then treated independently. We have taken this approach in our research, concentrating on heat flow by conduction, which has enabled us to make significant progress towards the goals mentioned above.

Model Overview

Our heat flow model is based on solving the heat conduction equation in cylindrical coordinates with a realistic source term using the method of Finite Differences (FD). Full details of this model can be found in our earlier publication [4]. The model incorporates temperature dependent material properties, accounts for the latent heat that accompanies melting and crystallization, can model time dependent deployment (where packages are emplaced over a period of time rather than all together), and is sufficiently detailed to provide both near and far field information.

Optimizing the SSM/Waste combination

Whilst HDSM is the preferred choice for a SSM, there are disposal scenarios in which it would be quite unsuitable. In what follows we outline how thermal modeling can be used to determine the conditions in which HDSM is optimum, and those for which an alternative SSM would be advisable. A more in-depth description of this modeling can be found in [5].

LTVDD 2 is particularly suited to the disposal of waste packages containing SNF from LWRs operating at burn-ups of 55 GWd/t or greater. Such waste could include spent MOX and/or uranium dioxide. The most efficient disposal scheme would entail the removal of the fuel pins from their assembly followed by emplacement in a waste container (consolidation) which is then in-filled with molten lead to remove all voids, and act as a partial radiation shield. Since there are several engineering 'degrees of freedom' e.g. number of fuel pins per waste package, age of the waste (post reactor), type of waste (UO_2/MOX and burn-up), number of waste packages per batch, it is not immediately obvious which combinations of these variables will lead to a thermal outcomes suitable for HDSM. Our heat conduction model solves this problem by enabling the construction of an optimization field diagram with boundaries demarcating those combinations of variables for which a given SSM is recommended.

FD modeling was used to determine the maximum temperatures reached at the outer surface of a single container of consolidated waste at various vertical distances from the borehole axis over the course of a simulation (covering the first 50,000 days following emplacement) (Figure 2). This figure shows that for the modeled combinations of age/waste type/number of pins, the HDSM in the annulus reaches a sufficiently high temperature over most of the height of the waste package. The maximum temperature is invariably reached near the centre of the package and can vary strongly depending on the contents.

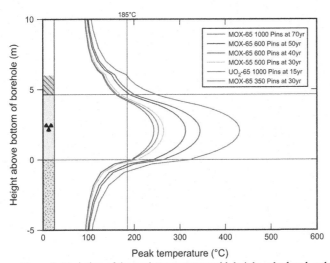

Figure 2. Variation of the peak temperature with height calculated at the surface of a single container of consolidated spent fuel pins (UO$_2$ and MOX at 55 and 65 GWd/t burn-ups). Vertical line indicates the temperature at which HDSM begins to melt. From left to right: MOX-65: 350 pins, UO$_2$-65: 1000 pins, MOX-55: 500 pins, MOX-65: 600 pins (50yr), MOX-65: 600 pins (40yr), MOX-65: 1000pins.

The peak temperature attained at the top, middle and bottom of a single waste package as a function of radial distance gives an indication of the thickness of the zone in which the HDSM would melt. Figure 3 shows the outcome from a single waste package containing 1000 pins of 15 year old spent UO$_2$ at 65 GWd/t burn-up. Figure 3 indicates that ambient temperature is reached about 25 m distance from the borehole axis.

A plot of temperature versus time at the outer surface of a waste container and at the borehole wall, enables the determination of the number of days taken to reach the onset of melting in the HDSM and amount of time needed before solidification begins. Figure 4 shows this scenario for the same waste package featured in Figure 3. By calculating temperature at points near the top, bottom and middle of the package, it is readily seen in this case that the HDSM is never melted at the top. In practice this would not be a serious problem because decay heat from the containers emplaced above the first package would provide the additional thermal input for a more geometrically complete melting.

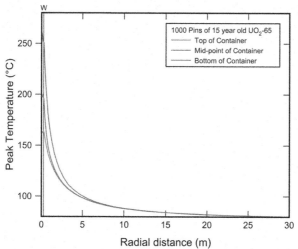

Figure 3. Peak temperature versus radial position at three vertical positions (top:lower curve, middle:upper curve and bottom) of a single container of spent UO₂ at 65 GWd/t burn-up.

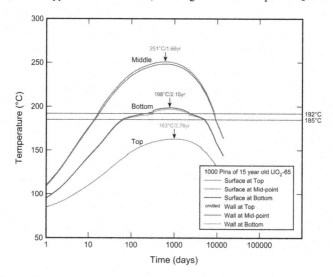

Figure 4. Variation of temperature with time calculated at three levels (top, middle and bottom) of a single container of consolidated spent UO₂ at 65 GWD/t burn-up at the outer edge of the waste package and at the borehole wall. Horizontal lines show the solidus and liquidus for HDSM.

The outcomes from a large number of modeling situations are conveniently summarized in the form of an optimization field diagram: Figure 5. With the aid of this diagram, it is possible to determine the best combinations of waste age and number of fuel pins for which HDSM can safely be used as the SSM.

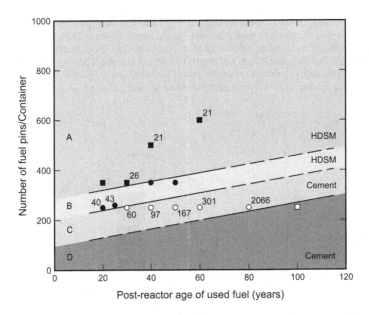

Figure 5. Outcomes of heat flow modeling for a single container of consolidated spent MOX at 65 GWd/t burn-up. The numbers are the number of days from initial disposal to the onset of melting in the HDSM. The lines serve to demarcate combinations of number of pins per container and age of the fuel for which the annotated SSM is a suitable choice.

For those combinations in which the heat output is insufficient to melt the HDSM adequately, an alternative SSM, such as a cementitious grout would be advised. Ideally, the thermal modeling work should be repeated using the thermophysical properties of an actual candidate cement to ensure that this is indeed a suitable SSM for those cases. We are currently developing a set of cements for DBD and plan to report the results of such heat flow modeling in the near future. A more interesting use of this thermal model would be to consider the outcomes obtained from disposing of mixed waste packages in a borehole. Various scenarios could be explored, from mixing fuel pins of different age, burn-up and type (UO_2 or MOX) within a single package to using packages of different content, say adding packages containing high burn-up MOX to a disposal of mostly (lower) heat output waste content (eg UO_2 or older MOX). These scenarios could greatly expand the flexibility of LTVDD 2 and other DBD concepts.

Sealing the borehole

The disturbed rock zone (DRZ) or engineering damage zone is a region of potential weakness in DBD. Conventional seals based on bentonite clay, asphalt, cement or swell packers for example, do not penetrate beyond the borehole itself. The same would be true of novel SSMs such as HDSM. A potentially superior sealing method involves "rock welding". Rock welding would entail backfilling the borehole with a slurry of crushed granite. A sacrificial heating device could then be buried into the slurry. Sufficient electrical power could then be supplied to the heater from the surface via a cable, to raise the temperature above the solidus – see Figure 6.

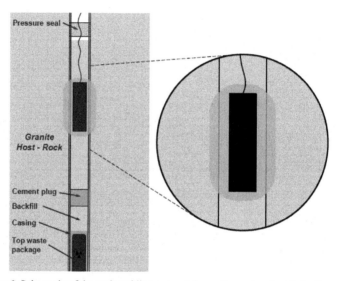

Figure 6. Schematic of the rock welding concept for sealing a deep borehole disposal.

The aim would be to melt not just the granite-water slurry but also the enclosing rock in and beyond the DRZ, so that the latter does not act as a bypass for the seal. Power to the heater could then be reduced or switched off after a predetermined time allowing the granite-melt to cool and recrystallize. Lab scale experiments demonstrate that granite recrystallized in this way is essentially indistinguishable from pristine samples when examined by optical microscopy (some difference in grain size is observed but this could be controlled through the cooling regime) [6].

Thermal modeling work is essential for confirming the feasibility of rock welding and for establishing the optimum operating parameters. Design of the electrical heating device is another area in which modeling is an essential input. Using the same FD approach described in the last section, we have modeled the case in which a simple cylindrical solid core heater is embedded in an annulus containing crushed rock. The electrical heater dimensions are: height 2m and diameter 0.172m, while the (constant) power output is taken to be 150 kWm^{-3}. We have taken the borehole diameter to be 0.43m. The temperatures recorded at the various mesh points during the course of the simulation are used to generate isotherms. These isotherms are extremely useful for indicating the extent and shape of the granite zone of melting (see Figure 7). For the case depicted in Figure 7, it is clear that the rock weld extends well beyond the borehole wall and into the EDZ. Current research in our laboratory is aimed at demonstrating recrystallization of granite at elevated pressure and temperature on larger samples than previously used coupled with modeling work aimed at optimizing the electrical heating device.

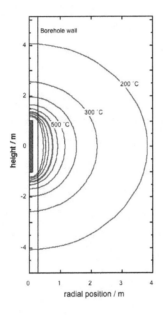

Figure 7. Maximum temperature isotherms at 100°C intervals generated for the modeled scenario involving a 2m long heater with a 0.172m diameter embedded in crushed rock within a 0.56m diameter borehole. The red coloured (thicker line) isotherm is the granite solidus. The power output of the heater is 150 kWm^{-3}.

CONCLUSIONS

Numerical modeling has played a key role in developing the concept of DBD. In this paper we have focused on two particular applications of heat flow modeling: optimization of the waste/SSM combination and a potentially superior method for sealing boreholes known as rock welding. Using finite differences, it has been possible to rapidly explore vast parameter spaces, and thereby greatly reducing the candidate systems to test when the first pilot borehole is constructed in 2016. Our modeling demonstrates that HDSM is a workable concept for an SSM, particularly when used for the disposal of higher burn-up spent fuel which would cause problems for more traditional sealing materials such as cementitious grout or bentonite. Future modeling work will be aimed at widening the envelope of wastes which may be disposed in a single hole while still using HDSM.

Numerical modeling has also validated the concept of sealing boreholes via rock welding. Isotherms generated by the models show that with sufficient power output, an electrical heating device could be designed which will result in a zone of melting which extends far out into the engineering disturbed zone and thus forming a secure, permanent seal. Thermal modeling is currently targeted at optimizing the design of the electrical heater prior to eventual trials in pilot borehole.

The above applications have involved models which have treated heat flow in isolation of any mechanical/hydraulic/geochemical processes. This is a convenient (though useful) simplification. Another application where this approach has proven fruitful is in determining the maximum vertical distance a particle might travel via convective flow. Because the timescales for convection and conduction are typically disparate, the problem may be solved in two parts; a solution for a point source of heat is first obtained and then used in a solution of the equations of fluid mechanics [4]. Models which incorporate a degree of explicit coupling between hydraulics/mechanics and heat flow do exist, but they are less well developed for DBD than they are for geologically shallow repository concepts.

REFERENCES

[1] A. J. Beswick, F. G. F. Gibb and K. P. Travis, "Deep borehole disposal of nuclear waste: engineering challenges", *Energy*, **167**, pp 47-66 (2014).
[2] NDA Report No. NDA/RWMD/083, "Geological Disposal: Review of options for accelerating implementation of the Geological Disposal programme", pp 31-37, (2011).
[3] F. G. F. Gibb, K. P. Travis, N. A. McTaggart, and D. Burley "High-density support matrices: - key to the deep borehole disposal of spent nuclear fuel", *J. Nuc. Materials*, 374, pp 370-377 (2007).
[4] F. G. F. Gibb, K. P. Travis, N. A. McTaggart, and D. Burley, "A model for heat flow in deep borehole disposals of high- level nuclear waste", *J. Geophys. Res. Solid Earth*, 113, B05201-B00508, (2008).
[5] F. G. F. Gibb and K. P. Travis, "Deep borehole disposal of higher burn up spent nuclear fuels", *Mineralogical Magazine*, **76**, 3003 (2013).
[6] P. G. Attrill and F. G. F. Gibb, "Partial melting and recrystallization of granite and their application to deep disposal of radioactive waste. Part 1 – rationale and partial melting.", Lithos

67(1-2), 103-117 (2003); P. G. Attrill and F. G. F. Gibb, "Partial melting and recrystallization of granite and their application to deep disposal of radioactive waste. Part 2 – recrystallization.", Lithos 67(1-2), 119-133 (2003).

Mater. Res. Soc. Symp. Proc. Vol. 1744 © 2015 Materials Research Society
DOI: 10.1557/opl.2015.314

Characteristics of Cementitious Paste for use in Deep Borehole Disposal of Spent Fuel and High Level Wasteforms

Nick C Collier[1], Karl P Travis[1], Fergus G F Gibb[1], Neil B Milestone[2]
[1] Immobilisation Science Laboratory, Department of Materials Science & Engineering, The University of Sheffield, Sheffield S1 3JD, United Kingdom.
[2] Callaghan Innovation, 69 Gracefield Road, PO Box 31310, Lower Hutt 5040, New Zealand.

ABSTRACT

Deep borehole disposal (or DBD) is now seen as a viable alternative to the (comparatively shallow) geologically repository concept for disposal of high level waste and spent nuclear fuel. Based on existing oil and geothermal well technologies, we report details of investigations into cementitious grouts as sealing/support matrices (SSMs) for waste disposal scenarios in the DBD process where temperatures at the waste package surface do not exceed ~190°C. Grouts based on Class G oil well cements, partially replaced with silica flour, are being developed, and the use of retarding admixtures is being investigated experimentally. Sodium gluconate appears to provide sufficient retardation and setting characteristics to be considered for this application and also provides an increase in grout fluidity. The quantity of sodium gluconate required in the grout to ensure fluidity for 4 hours at 90, 120 and 140°C is 0.05, 0.25 and 0.25 % by weight of cement respectively. A phosphonate admixture only appears to provide desirable retardation properties at 90°C. The presence of either retarder does not affect the composition of the hardened cement paste over 14 days curing and the phases formed are durable under conditions of high temperature and pressure.

INTRODUCTION

Deep borehole disposal (DBD) of spent nuclear fuel (SF) and high level waste (HLW) is seen as a viable alternative to emplacement in geologically shallow repositories [1, 2]. The process is based on locating and sealing waste packages at the bottom of boreholes drilled several kilometers into basement rock [2]. Building on pioneering work on deep boreholes over the past 20 years [3, and references therein], the DBD research group at The University of Sheffield in the UK is not only developing a "rock welding" technique for borehole sealing, but is also investigating the use of cementitious grouts for sealing and supporting emplaced waste packages when temperatures at the container surface do not exceed ~190°C [2]. The elevated temperature (and pressure) will accelerate grout thickening and setting time and will affect hardened paste properties [4, 5].

In this work we have drawn on knowledge from oil and geothermal well cementing applications. Grout thickening/setting time needs to be retarded to facilitate grout flow around the waste packages, after which the grout must set and develop sufficient strength before emplacement of subsequent packages. In proof-of-concept studies, grouting systems based on Portland cement are being investigated at small scale. Rheological properties and setting characteristics are studied at elevated temperature and pressure, and early age grout composition is investigated. Retarders are being examined to provide desirable thickening/setting properties. The work described here focuses on the influence of two organic retarders on grout properties.

EXPERIMENT

Class G oil well cement [6] partially replaced with silica flour was used to make the grout. The cement contained 55.11 wt% Ca_3SiO_5, 19.18 wt% Ca_2SiO_4, 1.03 wt% $Ca_3Al_2O_6$ and 16.25 wt% $Ca_4Al_2Fe_2O_{10}$, and the silica flour contained 98.56 wt% SiO_2. To enable flow of the grout through water, an underwater admixture (UCS Pak) supplied by Sika Ltd was used. The retarders assessed were sodium gluconate and Sika Retarder (a proprietary phosphonate product). Tap water was used to mix the grout and grout density (excluding retarder) was 1.892 kg/m^3. Grout consistency was investigated using a high pressure, high temperature, consistometer operating at 90, 120 and 140°C and 50 MPa with a linear 4 hour heating regime. For each grout that took longer than 4 hours to reach the required consistency, initial and final set was checked after 24 hours using manual Vicat testing equipment. Cement hydration reactions were investigated in an isothermal calorimeter set at 120°C and flow characteristics were investigated at ambient temperature using a modified ASTM flow cone test [7] to record time to flow. After curing at 120°C for 3, 7 and 14 days, grout composition was investigated using X-ray diffraction (XRD) and thermogravimetric analysis/derivative thermogravimetric analysis (TGA/DTG).

RESULTS

Times at which changes in grout consistency occurred were measured; t_1 was the time at which minimum consistency occurred and t_2 was the time for consistency to reach 70 Bearden units (Bc). 70 Bc is reported as the limit of pumpability (LoP) in well cementing applications [8], and has been taken as the consistency at which grout is likely to cease to flow around waste containers. Consistency plots for grouts containing sodium gluconate are shown in Figure 1, and the times resulting from all tests are summarized in Table 1. A retarder addition to give t_2 greater than 4 hours was desired; this target time is based on the time drillers are confident that a grout package could be delivered to a depth of 5 km, but options for faster times are being explored. The upper dosage level of Sika Retarder was based on the manufacturer's recommendation.

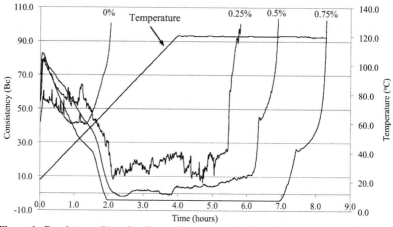

Figure 1: Consistency Plots for Grouts Containing Sodium Gluconate at 120°C

Addition Level (% BWOC)	Grout System (t_1, t_2 in hours)		
	90°C	120°C	140°C
	Unretarded		
0.0		1.0, 1.9	
	Sodium Gluconate		
0.025	1.6, 2.4	-	-
0.05	3.0, 4.1	-	-
0.075	3.0, 3.7	-	-
0.1	3.2, 10.0	-	-
0.25	-	2.1, 5.5	2.1, 4.0
0.5	-	2.3, 6.8	1.9, 4.4
0.75	-	1.9, 8.2	1.7, 5.2
1.0	-	-	1.7, 6.0
	Sika Retarder		
0.5	3.4, 5.7	-	-
1.0	3.3, 8.0	2.4, 3.2	-
1.5	-	2.5, 3.5	-
2.0	-	2.5, 3.4	-
3.0	-	2.3, 3.4	-

The progression of consistency for all grouts was similar showing an early increase in consistency followed by a decrease towards 0 Bc (which equates to 0.00782 Nm torque, and indicates a very low viscosity) and finishing with a final increase in consistency prior to setting. The unretarded system reached t_2 in less than 4 hours, demonstrating the need for retardation. To achieve t_2 greater than 4 hours at 90, 120 and 140°C, sodium gluconate addition levels of at least 0.05, 0.25 and 0.25 % by weight of cement (BWOC) respectively were required. Apart from t_2 being less than 4 hours for the 90°C sample containing 0.075 % addition, the general trend using sodium gluconate was that t_2 increased with addition level, and this additive provided the desired retardation at each temperature. Using Sika Retarder, only the grout tested at 90°C could be retarded sufficiently to give t_2 greater than 4 hours (at ≥ 0.5% BWOC). It was not possible to produce a sufficiently retarded grout at 120°C containing Sika Retarder using the addition levels studied here, so no testing was performed at 140°C. All grouts where t_2 was greater than 4 hours achieved initial and final set by 24 hours.

Because samples were mixed externally to the calorimeter, heat flow events that occurred before the samples heated up were not recorded. Adding increasing amounts of retarder changed the heat generation characteristics of the grout, especially with sodium gluconate. Even though it was difficult to identify any consistent trend in the magnitude of the main heat flow event in relation to retarder addition level, adding 0.5, 0.75 and 1.0 % BWOC sodium gluconate delayed the time at which the main heat flow event occurred to 9.0, 10.1 and 18.3 hours respectively after mixing. Sika Retarder was less effective than sodium gluconate even when larger quantities were added with 0.5, 1.0 and 2.0 % additions BWOC only causing retardation of the main heat flow event to 2.9, 3.2 and 4.9 hours respectively after mixing.

The flow results showed that as well as acting as a thickening/setting retarder, sodium gluconate also acted as a plasticiser with the time for the grout to flow through the testing cone reducing as the amount of sodium gluconate added increased from 0 to 1.0 wt% BWOC, with the shortest flow time at approximately 0.7 %. When the amount of Sika Retarder added was increased incrementally from 0 to 1.0 %, the time for the grout to flow through the testing cone increased, with the maximum time occurring at an addition level of approximately 0.75 %. These results show that water content may be reduced in the grout containing sodium gluconate in order to produce a less permeable matrix and provide better sealing of waste packages against access by saline groundwater. Any grout with t_2 greater than 4 hours was poured through water held at 63°C and its dispersion observed. Generally all these grouts flowed through the water well with only a small amount of solid particle dispersion. Any solid particles that did disperse soon settled on the top surface of the undispersed grout.

A summary of the phases detected in the hardened cement pastes is given in Table 2.

Table 2: Summary of Main Phases Detected in all Samples of Hardened Cement Paste using XRD and TGA/DTG

Sample	C-S-H	P	Quartz	α-C$_2$SH	T-11	B	C
Control Grout Without Retarder							
3 days			✓✓✓✓✓	✓✓	-	✓	✓
7 days	Incr.	Decr.	✓✓✓✓	✓	✓	✓	✓
14 days			✓✓✓✓	-	✓	✓	✓
Grout Containing Sodium Gluconate							
3 days			✓✓✓✓✓	✓✓	-	✓	✓
7 days	Incr.	Decr.	✓✓✓✓	✓	✓	✓	✓
14 days			✓✓✓✓	-	✓	✓	✓
Grout Containing Sika Retarder							
3 days			✓✓✓✓✓	✓✓	-	✓	✓
7 days	Incr.	Decr.	✓✓✓✓	✓	✓	✓	✓
14 days			✓✓✓✓	-	✓	✓	✓

Notes:
1. P = Portlandite (Ca(OH)$_2$); α-C$_2$SH = Ca$_2$(SiO$_4$)H$_2$O; T-11 = tobermorite-11Å (Ca$_5$Si$_6$(OH)$_{18}$.5(H$_2$O)); B = brownmillerite (Ca$_2$FeAlO$_5$); C = calcium carbonate (CaCO$_3$).
2. The number of ✓'s represents relative quantities of phases detected within each separate sample based on the intensity of the main XRD reflections for each phase.

The main crystalline phase in all samples was quartz (SiO$_2$) from the silica flour. Three forms of calcium silicate hydrate were identified; α-C$_2$SH (Ca$_2$(SiO$_4$)H$_2$O) and Tobermorite-11Å (Ca$_5$Si$_6$(OH)$_{18}$.5(H$_2$O)) were detected by XRD, and an amorphous material, usually referred to as C-S-H [9], which was responsible for a hump centered at approximately 37° 2θ in all XRD traces. Crystalline portlandite (Ca(OH)$_2$) was not detected by XRD but the TGA/DTG results identified this phase in all samples suggesting the presence of amorphous portlandite [10], and carbonation of this material was evident by the detection of calcite. A small quantity of brownmillerite (Ca$_2$FeAlO$_5$) was detected in each sample. There was no detectable difference in the compositions of the hardened cement pastes due to the presence of either retarder. For each grout, the intensity of the quartz XRD reflections reduced slightly between 3 and 7 days suggesting quartz reaction. α-C$_2$SH was identified in all samples cured for 3 and 7 days with the intensities of the main reflections reducing with time so by 14 days it was difficult to detect. Tobermorite-11Å was not detected in any sample at 3 days curing, but small XRD reflections were identified in all samples cured for 7 and 14 days (it was difficult to identify one of the strongest reflections for this phase which should have been at 9.2° 2θ, but the other main reflections were identified). The quantities of the calcium silicate hydrate phases increased with time whilst the amount of portlandite decreased. These progressive phase changes follow those suggested in the literature with 1) quartz slowly reacting with calcium to form calcium silicate hydrate phases, 2) the early formation of α-C_2SH, the quantity of which gradually reduces with time, and 3) the formation of tobermorite-11Å following the decline of α-C_2SH [8, 11]. Although not presented here, TGA/DTG data showed that the quantities of the calcium silicate hydrate phases in samples containing either retarder were less than those formed in the control samples.

DISCUSSION

Sodium gluconate and Sika Retarder delay the onset of thickening and setting, with the former having the greater influence. Further investigations are required to identify more accurate addition levels at all temperatures, and to demonstrate the repeatability of the consistency results. Addition of sodium gluconate also reduces the time of the grout to flow through the testing cone at ambient temperature and pressure, which indicates that it is also acting as a plasticiser. Of the two retarders studied, only sodium gluconate offered the potential to provide the amount of retardation required for the DBD process at all testing temperatures. However, consideration should be given to the use of separate retarder and plasticiser components, so Sika Retarder should not be discounted. Although the consistency and calorimetry data indicate that both products provide retardation of grout thickening, it is difficult to correlate the two sets of results because of the different operating conditions used. Consistency testing best replicates application of the fresh grout because of the slow heating rate from ambient temperature and increase of pressure to conditions representative of those in the borehole.

The grout flow data obtained at ambient temperature is useful in comparing the performance of the different types of grout, but ideally this data needs to be obtained at temperatures and pressures representative of those down the borehole. Additionally, the overall sealing of waste containers in grout should be studied at temperatures and pressures representative of those in an actual disposal. The applicability of using the LoP (70 Bc) as the DBD limit for the ability of the grout to flow around waste containers can also only be assessed by carrying out waste encapsulation tests in conditions representative of those down a borehole.

The phases formed in the grouts are typical of those found in oil and geothermal well cementing applications, which suggests a high level of durability for the grouting system being developed. Additionally, the fact that all samples with $t_2 \geq$ 4 hours achieved final set by 24 hours provides confidence that waste deployment rates of the order of 1 package/day could be achievable.

The design of the grout mixing process used for DBD (likely to take place at the well head) will require consideration. Even though the major oil well standards [6] stipulate the use of vertical blender mixers like that used here, the most common grout mixing systems used in the field are based on jet mixing processes [8]. Assessment of the applicability of using jet mixing processes and their effect on grout rheology will be required and may be complex because of the differences in grout rheology when mixing and deploying at different temperatures and pressures. Low permeability of the hardened grout will be required to provide sealing of the waste packages which means a low grout water content and the use of superplasticisers. However, this will also be governed by the ability to mix the grout at ambient conditions.

Increasing temperature accelerates the onset of grout thickening [12, 13]. Although a similar effect is reportedly caused by elevated pressure, this is not expected to be as significant as that of elevated temperature [4]. Similarly, deployment of the grout to the bottom of the borehole will require significant investigation, and should include assessment of whether grout cooling or stirring during deployment may be required.

CONCLUSIONS

Sodium gluconate appears to offer the potential to provide the amount of retardation required for use in DBD applications and also provides an increase in grout fluidity. The quantities of sodium gluconate required to be added to the grout to ensure desired fluidity for 4 hours at 90, 120 and 140°C are 0.05, 0.25 and 0.25 % BWOC respectively. At addition levels recommended by the manufacturer, Sika Retarder only provides desirable retardation properties at 90°C and does not influence fluidity advantageously. Neither retarder affects the composition of the hardened cement paste up to 14 days curing when compared to grouts without retarder. All phases detected in the hardened cement pastes are in line with those reported in the literature, which have demonstrated durability at high temperature and high pressure. All grouts where retardation was greater than 4 hours achieve initial and final set within 24 hours.

Further work is required to 1) enable reduction of grout water content and assess permeability, 2) consider other products available to retard consistency and setting, and 3) investigate other novel cementing systems that may be applicable to this application.

ACKNOWLEDGMENTS

The authors are grateful to the UK Engineering and Physical Science Research Council for funding (grant number EP/K039350/1).

REFERENCES

1. B. Arnold, P. Brady, S. Altman, P. Vaughn, D. Nielson, J. Lee, F. Gibb, P. Mariner, K. Travis, W. Halsey, J. Beswick, J. Tillman, FCRD-USED-2013-000409, SAND2013-9490P, Sandia National Laboratories report for U.S. Department of Energy, October 25, 2013.
2. J. Beswick, F. Gibb, K. Travis, Proceedings of the ICE - Energy, 167 (2014) 47–66.
3. F. Gibb, N. McTaggart, K. Travis, D. Burley, K. Hesketh, J. Nucl. Mat. 374 (2008) 370–377.
4. G. Scherer, G. Funkhouser, S. Peethamparan, Cem. Concr. Res. 40 (2010) 845–850.
5. A. Jupe, A. Wilkinson, K. Luke, G. Funkhouser, Cem. Concr. Res. 38 (2008) 660–666.
6. BS EN ISO 10426-1:2009, British Standard Institute.
7. ASTM C939-10, American Society for Testing and Materials.
8. E. Nelson, D. Guillot (Eds.), Well Cementing, 2nd Edition, Schlumberger, USA, 2006.
9. H. Taylor, Cement Chemistry, 2nd Edition, Thomas Telford, London, 1997.
10. H. Midgley, Cem. Concr. Res 9 (1979) 77-82.
11. J. Bensted, Development with Oilwell Cements, in Structure and Performance of Cements, J. Bensted, P. Barnes (Eds), 2nd Edition, Spon Press, London, 2008.
12. A. Shariar, M. Nehdi, Proceedings of the ICE – Construction Materials, 165 (2012) 25-44.
13. J. Zhang, E. Weissinger, S. Peethamparan, G. Scherer, Cem. Concr. Res. 40 (2010) 1023–1033.

Mater. Res. Soc. Symp. Proc. Vol. 1744 © 2015 Materials Research Society
DOI: 10.1557/opl.2015.474

Physicochemical Properties of Vitrified Forms for LILW Generated From Korean Nuclear Power Plant

Cheon-Woo Kim, Hyehyun Lee, In-Sun Jang, Hyun-Jun Jo, Hyun-Je Cho
Central Research Institute of KHNP, 1312-70 Yuseong-daero, Yuseong-gu, Daejeon 305-343,
Republic of Korea

ABSTRACT

Since 1994, the KHNP has developed a vitrification technology to treat the LILW generated from Korean nuclear power plant. To vitrify the LILW including combustible Dry Active Waste (DAW) and Ion Exchange Resin (IER) containing Zeolite, two borosilicate glasses are formulated. One of the formulated glass, DG2, is for the DAW vitrification solely and the other one, AG8W1, is for the blended wastes (DAW & IER) vitrification in a commercial vitrification facility in HanUl (former Ulchin) nuclear power plant. The physicochemical properties of the two glasses have been evaluated. To evaluate the processability of the glasses, the viscosities and electrical conductivities of the glass melts were measured in the laboratory within a temperature range between 950 and 1,350 degrees C, respectively. The liquidus temperatures of the glasses were evaluated using a gradient furnace for DG2 and data from heat treatment for AG8W1. The Mössbauer spectroscopy for AG8W1 was employed to evaluate the relations between the redox equilibria of iron. In addition, to verify the waste acceptance criteria for the final disposal of the vitrified forms, the compressive strengths of the vitrified forms were tested after an immersion test, a thermal cycling test, and an irradiation test. To verify the chemical durability of the glasses, several tests such as PCT, ISO, VHT, Soxhlet, MCC-1, and ANS16.1 were carried out. The PCT showed leach rates of B, Na, Li and Si were much less than those of the benchmark glass. The ISO test was performed at 90 degrees C for 1,022 days and Cumulative Fraction Leached of all elements in the glasses were analyzed. According to the VHT, the glasses had an outstanding chemical resistance under humid environment at 200 degrees C for 7 days. The Soxhlet leaching was performed on rectangular glass samples at 98 degrees C for 30 days. To analyze the forward dissolution rates of major glass elements, the MCC-1 was conducted at temperatures of 40, 70, and 90 degrees C for three weeks in pH buffer solutions ranging from pH 4 to 11. The processability of the glasses was in the desired ranges. And the product quality of the glasses met all regulatory guidelines. Using two glasses, the CCIM commissioning tests in the UVF were successfully performed and they showed good workability.

INTRODUCTION

The KHNP has investigated and evaluated various efficient thermal treatment technologies for the low-and intermediate-level radioactive waste (LILW) that is generated from nuclear power plants (NPPs). The KHNP considered vitrification technology to be a candidate for the treatment of the LILW generated from NPPs. The UVF project was launched in September 2002 with governmental support. The final operation permit was acquired in October 2009 and this initiated the commercial operation. The target waste streams for the UVF are combustible dry active waste (DAW) and W1 waste which is a blend of DAW and IER. The

UVF was designed to have ~18-20 kg/h of normal waste feeding capacity into the CCIM(cold crucible induction melter). A simplified arrangement of the UVF process is presented in figure 1.

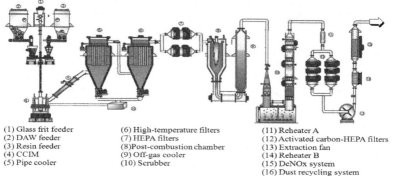

(1) Glass frit feeder	(6) High-temperature filters	(11) Reheater A
(2) DAW feeder	(7) HEPA filters	(12) Activated carbon-HEPA filters
(3) Resin feeder	(8) Post-combustion chamber	(13) Extraction fan
(4) CCIM	(9) Off-gas cooler	(14) Reheater B
(5) Pipe cooler	(10) Scrubber	(15) DeNOx system
		(16) Dust recycling system

Figure 1. Schematic diagram of main system of UVF.

To vitrify DAW and W1 in the UVF, two borosilicate glasses were formulated as shown in table 1. The formulated glass DG2 is solely for DAW vitrification and the other formulated glass AG8W1 for the blend of DAW and IER. To evaluate the physicochemical properties of the two glasses, several property measurements were performed and their results have been discussed in this paper.

Table 1. Compositions of DG2 and AG8W1 glasses (in wt.%).

	Al_2O_3	B_2O_3	CaO	Fe_2O_3	K_2O	Li_2O	MgO	Na_2O	SiO_2	TiO_2	ZrO_2	Others
DG2	7.07	11.30	9.77	0.35	4.47	5.25	4.63	10.07	41.25	3.09	1.13	1.62
AG8W1	12.30	9.97	4.82	1.78	1.63	1.24	2.12	17.57	43.14	1.24	0.93	3.26

EXPERIMENT

Physical properties measurement

The viscosity of the glass melt was obtained from measurement of the torque at various rotational speeds using a spindle attached to a viscometer. The viscosity of the glass melt was measured within a temperature range of 950 to 1,350 degrees C. The measured data was interpolated to temperature using the Vogel-Fulcher-Tammann(VFT) equation: $\ln \eta = A + B/(T - T_0)$, where A, B, and T_0 are fitting parameters. The electrical conductivity of the glass melt was determined by measuring the resistance of the glass melt as a function of frequency using a Pt/Rh electrode probe attached to an impedance analyzer. The electrical conductivity of the glass was measured in the temperature range of 950 to 1,350 degrees C. The electrical conductivity data was interpolated to temperature using the Arrhenius equation: $\log \sigma = A/T + B$, where A and B are fitting parameters. Scanning Electron Microscope(SEM) coupled with Energy Dispersive Spectrometer(EDS) was used for microstructural evaluation in this study. Samples of glasses were prepared for observation in a JSM-5600. The SEM was equipped with an Inca energy dispersive x-ray spectrometer. A gradient furnace and a heat treatment were used to identify the liquidus temperatures of DG2 and AG8W1, respectively. A Mössbauer spectrometer with a moving source of Co-57 in a Rh matrix was employed to evaluate relations between the redox equilibria of iron. The Fe^{2+} to total iron ratio was determined from the area of the Fe^{2+} peak divided by the sum of the combined areas of the Fe^{2+} and Fe^{3+} peaks. The purpose of the compressive strength test is to evaluate physical durability of a vitrified form. They are used as means for comparing compressive strengths before and after the immersion test (90 days at room temp.), the thermal cycling test (-40~60 degrees C for 30 times) and the irradiation (10^7 Gy of γ-ray) test. Three each of test specimens for a vitrified form have been prepared in a cylindrical form of 50mm in diameter, 100mm in length. Compressive strength for a vitrified form should be greater than 3.44 MPa (500 psi).

Chemical properties measurement

The Product Consistency Test(PCT) was used to evaluate the relative chemical durability of the glass by measuring the concentrations of the chemical species released from crushed glass powder (75~149 μm) in a 304L stainless steel vessel. A glass surface area to solution volume ratio is about 2,000m^{-1}. The test was conducted in deionized water at 90 degrees C for 7, 30, 60, and 120 days. The International Organization for Standardization(ISO) test was performed at 90 degrees C for 1,022 days and cumulative fraction leached (CFL) of major elements in the glass were evaluated. For the Vapor Hydration Test(VHT), the specimen (~5x~5x~10 mm^3) was suspended by a Teflon thread connected to a stainless steel support. The support along with the specimen was placed inside a Parr 22 ml T304 stainless vessel together with 0.25 ml of deionized water. The sealed vessel containing the sample was heated at 200 degrees C for 7 days. After removing the sample from the vessel, the sample was cross-sectioned and polished. The cross-section of the sample was examined and the thickness of the uncorroded part of the sample was measured by an optical microscope. The corrosion rate (g/m^2/day) of the specimen was then calculated. For the Soxhlet test, the temperature of water in the overflow vessel containing the glass sample was 98 degrees C and the rate of distilled water addition was approximately 200 ml/h. For the determination of the weight loss of the glass during the test time, the glass sample was taken out of the equipment after 1, 3, 6, 10, 17 and 30 days. The glass sample was dried and the weight loss of the glass was then determined. Five buffer solutions ranging from pH 4~11 were used for glass leachant for the Materials Characterization Center-1(MCC-1) test. The various pH values of the leachant are then measured before and after the leaching test at 40, 70, and 90 degrees C. The leachate of each glass specimen was analyzed using Inductively Coupled plasma-Mass Spectrometer(ICP-MS) for major glass components (B, Na, Al, Si, Co, Cs). The American Nuclear Society 16.1(ANS16.1) test was performed for 90 days after spiking three simulated elements, and leachability indices of the elements Cs, Co, Sr should be higher than 6.

DISCUSSION

Physical properties

The viscosities and electrical conductivities of DG2 and AG8W1 glass melts were measured as 9.87 and 42.8 dPa•s, and 0.26 and 0.30 S/cm, respectively at the processing temperature, 1,150 degrees C. It was evaluated that those parameters were well within the acceptable ranges (<100 dPa·s and >0.1 S/cm) for the proposed method of processing [1,2]. The activation energies were determined to be 135 and 152 kJ/mol for the viscosities of DG2 and AG8W1, respectively, 77.7 and 70.46 kJ/mol for the electrical conductivities of DG2 and AG8W1, respectively within a temperature range of 950 to 1,350 degrees C. The activation energies were similar to those of other borosilicate melts[3-4]. But the activation energies of the viscosity and the electrical conductivity were considerably higher than the reported values of about 70 and about 20 kJ/mol, respectively, for iron phosphate glasses [5].

The liquidus temperature of DG2 using a gradient furnace was evaluated to be about 864 degrees C. A crystal growth was found by a polarization microscope. Using SEM/EDS, the crystal found in DG2 was analyzed to be a silicate high in Si and Ca with a small amount of Al, Mg, and Na. The liquidus temperature of AG8W1 was determined to be about 850 degrees C.

An Fe^{2+} fraction of 4 to 60% should provide a sufficient level of safety against foaming or precipitations and should be the goal for processing radioactive waste glass. Measured Fe^{2+} fraction in AG8W1 solely using Mössbauer spectroscopy, 42%, represents the glass melt that exists at an optimum condition [6-7].

The compressive strength test results have been observed to be greater than 16,000 and 17,000psi for DG2 and AG8W1, respectively, in far excess of the requirement value of 500psi. In addition, the results of the immersion test, the thermal cycling test and the irradiation test have been evaluated to give compressive strengths greater than 16,000 psi, sufficiently satisfying the requirement value of 500psi as shown in table 2.

Table 2. Compressive strengths after immersion, thermal cycling and irradiation

Tests	DG2			AG8W1		
	Test #1	Test #2	Test #3	Test #1	Test #2	Test #3
Immersion	19,723	18,561	18,344	17,707	18,555	16,120
Thermal Cycling	18,858	18,741	18,812	18,696	18,473	18,835
Irradiation	18,968	18,892	18,640	18,988	18,773	18,911

Chemical properties

As shown in table 3, leach rates of B, Na, Si and Li evaluated from the PCT results for DG2 and AG8W1 were much less than those of the benchmark glass (SRL-EA) which is the reference glass for high-level vitrification in the USA and well below the current US DOE specification, 2g/m^2 for a 7-day test, for Hanford LAW borosilicate glass [8].

Table 3. Leach rate (g/m^2) of B,Na,Si,Li in DG2, AG8W1 and SRL-EA.

	7day			30day			60day			120day		
	DG2	AG8W1	EA	DG2	AG8W1	EA	DG2	AG8W1	EA	DG2	AG8W1	EA
B	0.93	0.85	9.16	1.35	1.23	6.84	1.20	1.74	7.74	1.80	1.32	21.0
Na	0.82	0.56	6.99	1.20	0.90	5.31	1.08	1.14	7.02	1.68	1.20	14.04
Si	0.05	0.21	2.12	0.27	0.24	2.34	0.24	0.24	2.28	0.24	0.24	3.36
Li	0.04	0.22	5.02	0.60	0.54	3.84	0.60	0.54	5.00	0.96	0.72	6.72

The cumulative fraction leached (CFL) of major elements in AG8W1 was solely analyzed using the ISO test. The elements Li, Na, and Si seemed to be saturated at a level below the CFL of 0.4, except that the leaching of B increases to give a non-linear relationship with time. Figure 2 shows the CFLs for several minor elements in AG8W1 after 1,022 days leaching at 90 degrees C. About 50 μm thickness composed of Si, Al, Fe, Ti, and Zr was found to be an altered layer. It has been assumed that the layer played a role as a prevention barrier against continuous corrosion and retarded the leaching to lower values in solution.

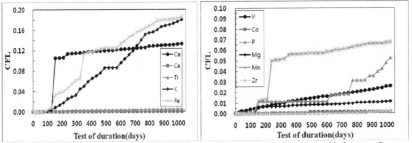

Figure 2. CFL for minor elements in AG8W1 after 1,022 days ISO test at 90 degrees C.

After the 7-day VHT at 200 degrees C, the corrosion rates of the glasses were evaluated as 10 and 2 g/m^2/day for DG2 and AG8W1, respectively and met the DOE specification (50 g/m^2/day) [5].

Based on the Soxhlet test, as shown in figure 3, the weight losses of DG2 and AG8W1 were determined as 322.4 and 106.8 g/m^2, respectively and were similar to and lower than those (max. 460 g/m^2) of other HLW glasses developed in Germany [9].

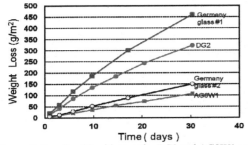

Figure 3. Results of Soxhlet test for DG2 and AG8W1 compared with Germany HLW glasses.

The forward dissolution rates of B, Na, Al, Si, Co, Cs in DG2 and AG8W1 were analyzed at each leaching condition. The test durations were 14, 21 and 91 days at 90, 70 and 40 degrees C, respectively. AG8W1 showed a forward dissolution rate of under 10 g/m^2•d when the temperatures were between 40 and 90 degrees C and the leachant condition was in a range of pH 4~11. Except for the DG2 glass, the minimum forward dissolution rate (0.01~1 g/m^2•d) was obtained at approximately pH 7~8. The forward dissolution rates of DG2 and AG8W1 were found to be V-shaped pH dependence as shown in the US DOE's research works [10-12].

The leachability indices of the major elements (Cs, Co, Sr) have been evaluated to be greater than 12 which satisfy being greater than 6 according to the ANS16.1 leaching test. Although the present ANS16.1 test method is a test method to evaluate leaching characteristics of a cement solidified waste, it is currently specified as a leaching test method for all solidified wastes in the criteria for accepting of radioactive wastes at a disposal facility in Korea, it should also apply to a vitrified form.

CONCLUSIONS

For the vitirification of the combustible DAW and blended waste (DAW and IER) generated from the Korean nuclear power plant, two borosilicate glasses were developed. DG2 is for vitrifying the DAW and AG8W1 for the blended waste. The process parameters such as viscosity and electrical conductivity of the glasses were well within acceptable ranges. The chemical property of the developed glasses met current requirements of the durability based on the results of PCT, VHT, ANS16.1, etc.

In addition, the compressive strength and leachability of the vitrified forms met the Korean waste acceptance criteria for the final disposal stability requirement. Using the cold crucible induction melter in the UVF, the workability and reliability of DG2 and AG8W1 glasses were successfully demonstrated and have contributed to the successful commissioning of the vitrification facility. The UVF is the world's first operating vitrification facility for the LILW associated with an NPP. The KHNP vitrification technology will not only enhance the safety of radioactive waste disposal but also promote further growth of the Korean nuclear power generation program.

REFERENCES

1. Jantzen, C. M., Ceramic Transactions, 23, 37-51 (1991).
2. Bickford, D.F., Ramsey, A. A., Jantzen, C.M., and Brown, K.G., J. Am. Ceram. Soc., 73(10), 2896-2902 (1990).
3. D.R. Uhlmann and N.J. Kreidl(Eds.), Glass Science and Technology, (Academic Press Inc., New York, 1986), pp. 233-270.
4. Doremus, R., Glass Science, 2nd ed., (John Wiley & Sons Inc., New York, 1994), Ch. 6 & 15.
5. Kim, C.W., and Day, D.E., J. Non-Cryst. Solids, 331, 20-31 (2003).
6. Schreiber, H.D., and Hockman, A.L., J. Am. Ceram. Soc., 70(8), 591-594 (1987).
7. Kim, C.W., Ray, C.S., Zhu, D., Day, D.E., Gombert, D., Aloy, A., Mogus-Milankovic, A. and Karabulut, M., J. Nuclear Materials, 322, 152-164 (2003).
8. US DOE, Design, Construction, Commissioning of the Hanford Tank Waste Treatment and Immobilization Plant, DOE Office of River Protection, Richland, WA, Contract No.: DE-AC27-01RV14136, (2001).
9. Personal communication with Siegfried Weisenburger at INE in Germany, (2005).
10. B.P McGrail et al., "Measurement of kinetic rate low parameter on a Na-Ca-Al borosilicate glass for low-activity waste," J. Nucl. Mater., 249, 175-189 (1997).
11. US DOE Argonne National Laboratory Report, ANL-05/43, 48-57 (2005).
12. US Department of Energy Report ANL–EBS-MD–000016, Rev. 00, ICN 01, Civilian Radioactive Waste Management System, Las Vegas, NV (2000).

Mater. Res. Soc. Symp. Proc. Vol. 1744 © 2015 Materials Research Society
DOI: 10.1557/opl.2015.372

Release of [108]Ag from Irradiated PWR Control Rod Absorbers under Deep Repository Conditions

O. Roth[1], M. Granfors[1], A. Puranen[1], K. Spahiu[2]
[1]Studsvik Nuclear AB, Hot Cell Laboratory, SE-611 82 Nyköping, Sweden
[2]SKB, Box 250, SE-101 24, Stockholm, Sweden.

ABSTRACT

In a future Swedish deep repository for spent nuclear fuel, irradiated control rods from PWR nuclear reactors are planned to be stored together with the spent fuel. The control rod absorber consists of an 80% Ag, 5% Cd, 15% In alloy with a steel cladding. Upon in-reactor irradiation [108]Ag is produced by neutron capture. Release of [108]Ag has been identified as a potential source term for release of radioactive substances from the deep repository.

Under reducing deep repository conditions, the Ag corrosion rate is however expected to be low which would imply that the release rate of [108]Ag should be low under these conditions. The aim of this study is to investigate the dissolution of PWR control rod absorber material under conditions relevant to a future deep repository for spent nuclear fuel. The experiments include tests using irradiated control rod absorber material from Ringhals 2, Sweden. Furthermore, un-irradiated control rod absorber alloy has been tested for comparison. The experiments indicate that the release of Ag from the alloy when exposed to water is strongly dependent on the redox conditions. Under aerated conditions Ag is released at a significant rate whereas no release could be measured after 133 days during leaching under H_2.

INTRODUCTION

In a nuclear reactor control rods are used to control the neutron flux and thereby the fission rate. In PWR reactors the control rods are grouped in assemblies of about 20 rodlets. The rodlets consist of an absorber alloy (80% Ag, 5% Cd, 15% In) surrounded by a cladding of stainless steel. PWR control rods are primarily used for startup and shutdown, while the power during an operating cycle is controlled by varying the boron content in the moderator and by burnable absorbers in the fuel. The control rod assemblies are inserted from the top of the core. Since the rods are normally retracted with only the bottom tip facing the core, the bottom end of the control rodlets will be more heavily irradiated than the top end.

In a future Swedish deep repository the PWR control rods are planned to be stored together with the fuel assemblies. Upon reactor operation, the composition of the control rod absorber will change due to neutron capture. The nuclides formed in these reactions may be released to ground water upon water intrusion in the deep repository. In a scenario where the barriers break at a relatively early stage (< 10 000 years), the release of [108]Ag has been identified as a nuclide that may be released from the deep repository [1]. In this risk assessment immediate release of [108]Ag is assumed although, the solubility of Ag is expected to be low under the reducing conditions expected to prevail in the deep repository [2]. However, to the best of our knowledge, no data on the release of [108]Ag from irradiated control rods have been published in the open literature.

In this study the release of Ag from irradiated and unirradiated control rod absorbers has been studied under various experimental conditions. Furthermore, the radionuclide inventory of irradiated control rod absorber material has been determined.

EXPERIMENT

Leaching of unirradiated control rod absorber

Three series of leaching of unirradiated material were performed, where the leaching solution and atmosphere were varied according to the following list:

I.   10 mM NaCl + 2 mM NaHCO$_3$, purging with Ar + 0.03% CO$_2$
II.   1 mM NaCl, purging with Ar + 0.03 % CO$_2$
III. 10 mM NaCl + 2 mM NaHCO$_3$, open to air

In each experiment 1 g of unirradiated Ag-Cd-In material (80% Ag, 15% In, 5% Cd) was immersed in 200 mL leaching solution in a glass vessel. The experiments were performed at room temperature. Series I and II were performed in a glovebox in order to ensure oxygen free conditions. The leaching was conducted for 250 days, during which 5 samples à 10 mL were withdrawn at different time intervals. The samples were analyzed with ICP-MS for Ag, Cd and In. The pH values of the leaching solutions were analyzed prior to and after the experiment.

Determination of nuclide inventory in irradiated control rod absorber

The examined control rodlet (rodlet F9) was chosen from the control rod R035 used in Ringhals 2. The control rod was manufactured by Westinghouse and belong to the original set of control rods from the start of the reactor. For the inventory determination, two 1 mm thick discs (F9-1 and F9-3) were cut from the rodlet (see Figure 1). In Figure 2 the gamma profile of another rodlet (R035/A4) is shown. The A4 rodlet is from the same control rod assembly as F9 and the irradiation history are similar, for this reason the gamma profile of A4 gives an indication of the irradiation level also for the rodlet used in this study (F9).

As the irradiation level varies significantly along the rod axis (as shown by the gamma profile in Figure 2), it is expected that also the radionuclide content varies in the irradiated material. In order to get a representative value for the leached sample (F9-2), the two samples used for inventory determination were taken at equidistant positions on both sides of the leached sample. The samples were weighed (0.66 and 0.72 g respectively) and dissolved in an acid mixture (12 mL supra pure 7 M nitric acid and 0.5 mL 22 M HF) after which the samples were analyzed for Ag, Cd, In and Sn using HPLC-ICP-MS.

Figure 1. Cutting scheme of rodlet R035/F9

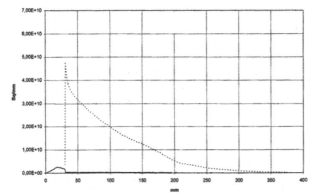

Figure 2. Gamma scan displaying ^{110}Ag distribution in rodlet R035/A4

<u>Autoclave leaching of irradiated control rod absorber</u>

The leaching of the irradiated material was performed in an autoclave pressurized to 4 bar with 10% H_2 and 0.03% CO_2 in Ar. The sample disc F9-2 (Figure 1) was cut in 4 pieces (total 1.98 g), placed in a glass basket and immersed in 400 mL 10 mM NaCl + 2 mM NaHCO$_3$ solution. The solution was purged with 0.03% CO_2 in Ar for about 2 hours prior to pressurization. The experiment was performed under stagnant conditions at room temperature. After 12 days of leaching, a gas sample was taken from the autoclave in order to ensure that no leakage occurred. Three liquid samples have been withdrawn from the autoclave at 12, 62 and 133 days of leaching. The liquid samples were acidified in order to avoid sorption on the vessel walls before analysis. The gas sample was analyzed with Gas-Mass Spectrometry and the liquid samples with gamma spectrometry (^{108}Ag) and ICP-MS (Cd, In and Sn isotopes). Isotopes of silver could not be determined by ICP-MS due to interference from silver lining on the autoclave seals.

RESULTS
<u>Leaching of unirradiated control rod absorber</u>

The results from the leaching of unirradiated control rod alloy are presented in Figure 3. As can be seen in the figure the release of Ag and Cd from the alloy is significantly higher during oxidizing conditions (i.e. under air) compared to the oxygen free leaching series. The release of In on the other hand seem to be more related to the HCO$_3^-$ concentration. In the samples were the leaching solution contained 2 mM NaHCO$_3$, the release of In was lower compared to the case where HCO$_3^-$ was introduced only by CO_2 purging. This can probably be explained by different pH in the solutions.

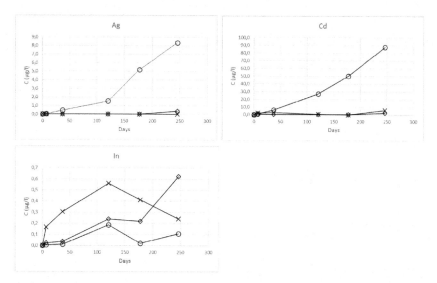

Figure 3. Release of Ag, Cd and In from leaching of unirradiated control rod alloy under varying experimental conditions. (◊) I. 10 mM NaCl + 2 mM NaHCO₃, (×) II. 1 mM NaCl, Ar + CO₂ Ar + CO₂, (o) III. 10 mM NaCl + 2 mM NaHCO₃, air

Determination of nuclide inventory in irradiated control rod absorber

During irradiation of the control rod material upon reactor operation the composition will change. The major neutron absorption reactions taking place are:

107 Ag + n	→	108 Ag	→	108 Cd (stable)		
109 Ag + n	→	110 Ag	→	110 Cd (stable)		
112 Cd + n	→	113 Cd (stable)				
113 Cd + n	→	114 Cd (stable)				
114 Cd + n	→	115 Cd	→	115 In (stable)		
115 Cd + n	→	116 Cd (stable)				
116 Cd + n	→	117 Cd	→	117 In	→	117 Sn
115 In + n	→	116 In	→	116 Sn (stable)		

The elemental and isotopic composition of the two samples from the irradiated control rod are shown in Figures 4-5. As can be seen in the figure, the largest change in composition compared to the nominal values are seen in sample F9-1. This is expected since this sample was taken closer to the bottom end of the control rod (i.e. the part closest to the reactor core) and hence has experienced a stronger radiation field during operation.

220

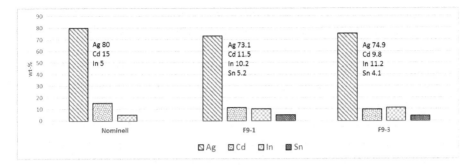

Figure 4. Composition of irradiated samples compared to nominal composition.

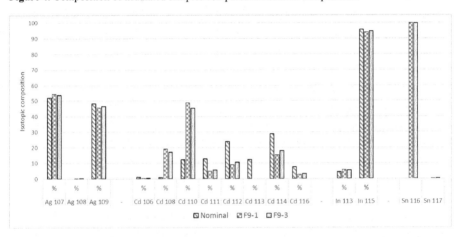

Figure 5. Isotopic composition of irradiated samples compared to nominal composition.

The isotopic composition shows that the radionuclide most relevant for the safety assessment of the future deep repository, ^{108}Ag, makes up only about 500 ppm of the total silver content. In sample F9-1 ^{108}Ag makes up 670 ppm of the total silver content and in sample F9-3 the fraction is 560 ppm. One must also keep in mind that these results are relevant for the part of the control rod closest to the reactor core, at distances further from the core the ^{108}Ag content will be even lower.

Autoclave leaching of irradiated control rod absorber

The gas sample taken from the autoclave at 12 days leaching showed no signs of air intrusion. The Cd, In and Sn content in the samples taken at 12 and 62 days leaching was all below detection limit of the ICP-MS (ppt range). The results from the analysis of ^{108}Ag by gamma spectrometry also showed results below the detection limit. For gamma analysis the detection limit is dependent on the measuring time. In this case the samples were measured for around 65 hours resulting in a detection limit below ppt levels. The third sample taken at 133 days leaching, also showed ^{108}Ag levels below detection limit (as above), whereas trace levels of Cd could be detected. The measured release of Cd^{2+} can be delayed by sorption on the autoclave inner walls.

DISCUSSION

From the results from the leaching of the unirradiated Ag/Cd/In alloy it is clear that the release of Ag and Cd is strongly dependent on the redox conditions. All three elements are oxidized by O_2. However, since $In(OH)_3$ is not soluble except under highly acidic conditions, the oxidation of In will not result in measurable amounts of In in solution. This explains why Ag and Cd is released at significant rates under aerated conditions, whereas the release of In seems more related to the pH of the solution. Since both Cl^- and HCO_3^- are present in the systems, the release of Ag and Cd is limited by the formation of AgCl(s) and $CdCO_3$(s) [3].

During the leaching of the irradiated material the conditions are initially reducing. Under these conditions, Ag should be stable in its metallic form and In and $In(OH)_3$. Cd however, can be released as Cd^{2+} unless the dissolution is limited by the formation of $CdCO_3$(s). The inherent radioactivity of the irradiated material will lead to the production of oxidizing water radiolysis products such as H_2O_2 and O_2. However, the results to date of the present study indicates that the inherent radiation of the alloy is not sufficient to cause significant dissolution of the material.

It should be noted that the radiation field in the present study originates from the alloy alone and is significantly weaker than the radiation field from the spent fuel which will dominate in a future deep repository. However, studies of oxidant production due to γ-radiolysis of water in presence of even lower dissolved hydrogen levels than our experimental conditions [4, 5], indicate that our conclusions about silver release should be valid even for higher radiation fields.

ACKNOWLEDGMENTS

Ringhals AB is gratefully acknowledged for providing the control rod absorber material and gamma scanning data.

REFERENCES

1. Long-term safety for the final repository for spent nuclear fuel at Forsmark, Main report of SR-Site project, Volume III, Svensk Kärnbränslehantering AB, Mars 2011
2. McNeil M B, Little B J, 1992. Corrosion mechanisms for copper and silver objects in near-surface environments. J. of American Institute for Conservation, 31, p. 355–366.
3. MEDUSA Chemical Equilibrium Diagrams, https://www.kth.se/en/che/medusa/downloads-1.386254
4. Pastina B, Isabey J, Hickel B, 1999. The influence of water chemistry on the radiolysis of primary coolant water in pressurized water reactors, J. Nucl. Mat., 264, pp. 309-318.
5. Pastina B, LaVerne J A, 2001. Effect of molecular hydrogen on hydrogen peroxide in water radiolysis, J. Phys. Chem. A., 105, pp. 9316-9322.

Mater. Res. Soc. Symp. Proc. Vol. 1744 © 2015 Materials Research Society
DOI: 10.1557/opl.2015.348

Advancing the Modelling Environment for the Safety Assessment of the Swedish LILW Repository at Forsmark

Henrik von Schenck[1], Ulrik Kautsky[1], Björn Gylling[1], Elena Abarca[2] and Jorge Molinero[2]
[1]Swedish Nuclear Fuel and Waste Management Co, SE-101 24 Stockholm, Sweden.
[2]Amphos 21 Consulting S.L., Passeig Garcia i Faria, 08019 Barcelona, Spain.

ABSTRACT

An extension of the Swedish final repository for short-lived radioactive waste (SFR) is planned and a safety assessment has been performed as part of the licensing process. Within this work, steps have been taken to advance the modelling environment to better integrate its individual parts. It is desirable that an integrating modelling environment provides the framework to set up and solve a consistent hierarchy of models on different scales. As a consequence, the consistent connection between software tools and models needs to be considered, related to the full assessment domain. It should also be possible to include the associated geometry and material descriptions, minimizing simplifications of conceptual understanding.

The usefulness of the analysis software Comsol Multiphysics as component of an integrating modelling environment has been tested. Here, we present two examples of hierarchical models. Consistent properties and boundary conditions have been extracted form regional hydrogeology and surface hydrology models when setting up repository scale models. CAD models of the repository have been imported into the analysis software, representing tunnel systems and storage vaults with engineered structures and barriers. Data from geographic information systems such as digital elevation models and geological formations have been also directly implemented into model geometries.

The repository scale hydrology models have provided a basis for further developments focussed on the modelling of coupled processes. An interface between Comsol Multiphysics and the geochemical simulator Phreeqc has been developed to support reactive solute transport studies. An important test case concerns radionuclide transport in a 3D, near-surface model of a catchment area. The dynamic surface hydrology has been simulated with MIKE SHE and connected to Comsol Multiphysics and Phreeqc for detailed hydro-geo-chemical modelling of radionuclide migration through soils and sediments.

INTRODUCTION

The Swedish final repository for short-lived radioactive waste (SFR) is located at the Forsmark site in the county of Uppland. The repository is situated in the bedrock, 50-120 m below the Baltic Sea. It holds low and intermediate-level operational waste from the Swedish nuclear power plants as well as industrial, research-related, and medical waste. An extension of the facility is planned, intended for future decommissioning waste. In figure 1, below, the existing facility is shown to the right and the planned extension to the left.

Figure 1 The Swedish repository for short lived radioactive waste at the Forsmark site. The existing facility is shown to the right and the planned extension to the left.

A safety assessment of both the existing facility and the planned extension has been performed as part of the licensing process. Within this work steps have been taken to improve the
- Consistency when linking different assessment models
- Level of detail in representing repository geometry and site data
- Ability to model coupled processes

The usefulness of the analysis software Comsol Multiphysics [1] as component of an integrating modelling environment has been tested and two examples are presented here to illustrate the development work. The first example concerns the modelling of groundwater flow through SFR. The second example concerns reactive transport in the sediments of the near-surface above the repository.

MODELLING EXAMPLES
Groundwater flow through SFR

The groundwater flow through the Forsmark site is modelled on a regional scale [2]. The hydrogeological model covers many square kilometres of land and reaches a depth of approximately one kilometre. In the SFR safety assessment, the software DarcyTools [3] has been used to set up and solve for the groundwater flow in the fractured granitic rock. The regional flow model accounts for the changing hydraulic boundary conditions resulting from the landscape evolution over time. The regional scale hydrology model includes the repository in its geometry representation. Nevertheless, a repository scale model has been considered to study the groundwater flow through the vaults and waste in greater detail and with greater flexibility [4].

The geometries of the SFR and its extension were available from planning and design work performed using the CAD-software Bentley MicroStation [5]. CAD geometries in the Parasolid format could be read into the Comsol Multiphysics analysis environment by means of the CAD Import Module [6], an add-on product to the basic software package. In Comsol Multiphysics, further work was performed to ready the geometry for analysis. This involved removing spikes, sliver face and detached surfaces from the geometry. For some repository vaults, engineered concrete structures were represented by adding geometry detail using the built-in geometry tools of Comsol Multiphysics. The analysis geometries of the SFR and the planned extension are shown in figure 2.

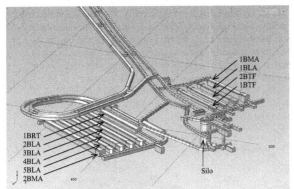

Figure 2 Analysis geometries of the SFR (right) and the planned extension (left). Indicated waste vaults include the silo for intermediate-level waste, 1-2BMA for intermediate level waste, 1-5 BLA for low-level waste and the BRT vault for reactor pressure vessels.

The repository scale models include surrounding rock volumes in addition to the repository geometry itself. In order to consistently link the repository scale model to the regional hydrology model the hydraulic property descriptions of the rock were transferred from DarcyTools to Comsol Multiphysics. This was accomplished by means of a user-programmed interface [4] able to read DarcyTools model files and write, for instance, the anisotropic hydraulic conductivity to a text file. Property fields could then be read into the near-field models and applied as property interpolation functions. The same interface also read pressure and flow velocity results from the regional model, to be applied as boundary conditions. This last step completed the consistent set up of the repository scale models, allowing groundwater flow simulations to be run.

Over the first thousand years of the assessment period it was found that the influence of land rise was the dominant process affecting groundwater flow through the repository [4]. Concrete degradation becomes important assessing the repository safety over thousands of years and beyond. It was found that degraded concrete barriers could continue protecting the waste from groundwater flow if surrounded by a high permeability backfill. In particular, simulations highlighted the importance of properly engineering the foundation of the barrier structures, to ensure high permeability also beneath the floor [7]. The main report for the safety assessment of the SFR outlines the evolution of the near-field hydrology in greater detail [8].

Reactive transport in the near-surface system

Should radionuclides escape from the repository they may be transported through the fractured rock and pass into the overlying regolith of the near-surface system. Analysing where and when radionuclides may reach the surface is an important task of the safety assessment. This requires accounting for groundwater flow, mass transport and geochemistry. The following example outlines a feasibility study aimed at modelling the fully coupled system. Focus was on increasing the level of detail describing the geometry, geological representation, hydrological behaviour as well as geochemical retention processes of the near-surface system.

The Forsmark site has been thoroughly investigated with respect to geographic data and landscape modelling has visualised the future development of the site [9]. The data are stored in the software ArcGIS [10] and describe the landscape topography, wetlands and lakes, the extent of soil layers and sediments, as well as the presence of engineered structures. In addition to the geometrical representation, available data also specify material properties, such as the porosity, permeability and sorption characteristics. This kind of coordinate based numerical data can be read into Comsol Mutiphysics, for instance, as parameterised surfaces. The geometry of a catchment area above the SFR repository is shown below. The landscape topography of the model is illustrated along with wetlands and a man-made pier (figure 3; left). The near-surface distribution of till, clay and rock is also shown (figure 3; right).

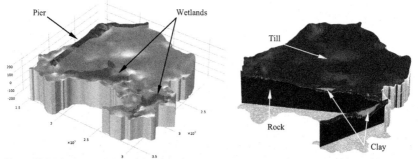

Figure 3 Model geometry and material representation of a future catchment area above SFR. The image to the left illustrates surface features including the landscape topography, wetlands and a man-made pier. The image to the right shows the considered geochemical domains of the near surface system, namely till, clay and rock.

Richards' equation was used to simulate groundwater flow in the variably saturated porous media using Comsol Multiphysics. Again, boundary conditions were extracted from an existing regional model of the near-surface hydrology [11] simulated in the MIKE SHE software [12], consistently integrating the local scale catchment model.

Species transport equations were also added in Comsol Multiphysics. The flow field contributes to advective mass transport and hydrodynamic dispersion. The transport equations are further affected by chemical reactions, effectively retarding the transport of radionuclides through the surface system. An existing geochemical model was available from previous work related to the Forsmark site [13]. The geochemical model describes interactions of selected radionuclides with the clay and till system, including precipitation/dissolution reactions, complexation to mineral surfaces, ion-exchange, as well as complexation with organic ligands. To add the geochemistry and complete the reactive transport model, an interface was programmed in java, communicating with Comsol Multiphysics by means of the available java API and calling the dynamic library iPhreeqc.dll [14]. In the resulting simulation environment

Comsol Multiphysics solved for the flow and coupled mass transport equations. In parallel, Phreeqc [15] solved for the local geochemical reactions.

With this set-up, transient reactive transport simulations were run for a set of radionuclides including Cs, Sr and U, entering the regolith from the bedrock below. The Cs concentration was found to decrease by five orders of magnitude passing from the bedrock to the surface. This was mainly due to the high affinity of Cs for the frayed edge sites of illite that retain Cs via cation exchange. Figure 4 shows the spatial distribution of Cs in the modelling domain after 1100 years, illustrating how dissolved species in the deep groundwater sorb onto different illite sites in clay and till.

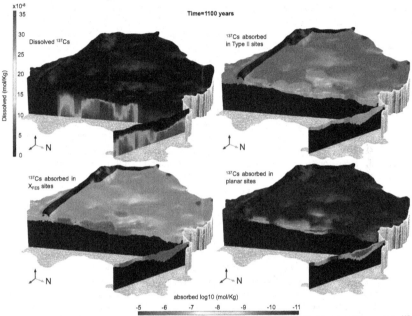

Figure 4 Top left: Predicted distribution of ^{137}Cs at the end of the simulation period (1100 years). Bottom left: ^{137}Cs retained in the frayed-edge sites of illite. Top right: ^{137}Cs retained in the type II sites of the illite interlayer. Bottom right: ^{137}Cs retained in the planar sites of the illite interlayer.

Sr was also retained somewhat by sorption to illite, with approximately 4% of the concentration in the deep groundwater reaching the discharge area. The retention of U in the regolith was found to be dominated by sorption to $Fe(OH)_3$ and by precipitation to form uraninite. In addition to these details, a main result from this study was the successful demonstration of the feasibility to model migration and retention of radionuclides in a realistic representation of the near-surface system above SFR. In this regard, the modelling approach serves as a basis for refining existing assessment models [16].

SUMMARY AND CONCLUSIONS

The planned extension of the existing Swedish repository for short-lived radioactive waste has motivated a review of the facility's safety assessment. As part of that work, steps have been taken to improve the consistency when linking different models and modelling tools. Other priorities have included improving the level of detail in representing repository geometry and site data, as well as the modelling of coupled processes.

The successful use of Comsol Multiphysics as a central component of the evolving modelling environment has been demonstrated. The software has been interfaced with the modelling tools DarcyTools and MIKE SHE to create repository scale models simulating groundwater flow, consistently linked to regional scale flow models. Comsol Multiphysics has also been able to import detailed representations of the SFR geometry and the near-surface system above the repository, interfacing with Bentley Microstation and ArcGIS software. Repository scale hydrology models have been further advanced by coupling to detailed geochemistry, interfacing Comsol Multiphysics with Phreeqc.

REFERENCES

1. Comsol Multiphysics v4.3 and v4.4, COMSOL AB, Stockholm, Sweden.
2. M. Odén, S. Follin, J. Öhman, P. Vidstrand, "SR-PSU Bedrock hydrogeology", R-13-25, Swedish Nuclear Fuel and Waste Management Co, Sweden (2014).
3. U. Svensson and M. Ferry, *Journal of Applied Mathematics and Physics* 2, 365 (2014).
4. E. Abarca, A. Idiart, L.M. de Vries, O. Silva, J. Molinero and H. von Schenck, "Flow modelling on the repository scale for the safety assessment SR-PSU", TR-13-08, Swedish Nuclear Fuel and Waste Management Co, Sweden (2013).
5. Bentley MicroStation v8i, Bentley Systems Inc, Exton, U.S.A.
6. Comsol CAD Import Module, COMSOL AB, Stockholm, Sweden.
7. E. Abarca, O. Silva, A. Idiart, A. Nardi, J. Font and J. Molinero, "Flow and transport modelling on the vault scale", R-14-14, Swedish Nuclear Fuel and Waste Management Co, Sweden (2013).
8. SKB, "Safety analysis for SFR Long-term safety. Main report for the safety assessment SR-PSU", TR-14-01, Swedish Nuclear Fuel and Waste Management Co, Sweden (2014).
9. T. Lindborg, L. Brydsten, G. Sohlenius, M. Strömgren, E. Andersson and A. Löfgren, *Ambio* 42, 402 (2013).
10. ArcGIS, Esri, Redlands, U.S.A.
11. S. Berglund, E Bosson and M. Sassner, *Ambio* 42, 425 (2013).
12. D.N. Graham and M.B. Butts, "Flexible, integrated watershed modelling with MIKE SHE", *Watershed Models*, ed. V.P. Singh and D.K. Frevert (CRC Press, 2005) pp. 245-272.
13. A. Piqué, D. Arcos, F. Grandia, J. Molinero, L. Duro and S. Berglund, *Ambio* 42, 476 (2013).
14. A. Nardi, A. Idiart, P. Trinchero, L. M. de Vries and J. Molinero, *Computers & Geosciences* 69, 10 (2014).
15. D. L. Parkhurst and C. A. J. Apello, "Description of Input and Examples for Phreeqc Version 3 - A Computer Program for Speciation, Batch-Reaction, One-Dimensional Transport, and Inverse Geochemical Calculations" *U.S. Geological Survey Techniques and Methods*, Book 6, chap. A43 (2013).
16. SKB, "Biosphere synthesis report for the safety assessment SR-PSU", TR-14-06, Swedish Nuclear Fuel and Waste Management Co, Sweden (2014).

Mater. Res. Soc. Symp. Proc. Vol. 1744 © 2015 Materials Research Society
DOI: 10.1557/opl.2015.366

Analysis of radionuclide migration with consideration of spatial and temporal change of migration parameters due to uplift and denudation

Taro Shimada, Seiji Takeda, Masayuki Mukai, Masahiro Munakata and Tadao Tanaka
Nuclear Safety Research Center, Japan Atomic Energy Agency, 2-4 Shirakata-shirane, Tokai-mura, Ibaraki, 3191195, JAPAN.

ABSTRACT

Integrated safety assessment methodology that analyzes radionuclide migration reflecting the spatial and temporal changes of disposal systems was developed for a geological disposal site with uplift and denudation, and then some case analyses for an assumed site were carried out. The combination of uniform uplift and denudation has the largest effect on the radionuclide migration because the ground water flow velocity increases with decreasing depth from the ground surface. In the case without denudation, tilted uplift has more effect than uniform uplift because flow velocity in tilted uplift increases with increasing hydraulic gradient. The long-term change of the geological structures including the uplift and denudation, the hydraulic conditions, and the recharge and outlet of the ground water around a candidate site should be carefully investigated to determine the appropriate the place, depth and layout of the repository.

INTRODUCTION

In long-term safety assessment of geological disposal system for high level radioactive wastes, it is necessary to evaluate the impact on the radionuclide migration where ground water flow and water composition are changed with decreasing depth of the repository from the ground surface by uplift and denudation. When the position of interest is up-warped mountain, a block of strike-slip fault or reverse dip-slip fault, the ground surface and geological layers in the entire area of interest are uplifted uniformly with maintaining the original shape of the ground surface (uniform uplift). On the other hand, when the position of interest is on a wing part of fold structure at the compression field, the elevation gradient of the ground surface and the geological layers in east-west or north-south direction will increase gradually over time (tilted uplift)[1]. The difference of the type of the uplifts may affect ground water hydraulics on and around disposal site, which may result in chemical phenomena within engineered barrier and along the migration path from the repository to the outlet of natural barrier. It is necessary to understand the quantitative effects of uplift and denudation to radionuclide migration, taking into account the linkage between hydraulics and chemicals.

When denudation is taken into consideration for long-term evaluation, it had been often assumed that the uplifted part at the surface was deleted by denudation to keep the current shape and elevation of ground surface. The depth of repository from the ground surface is decreased while the shape and elevation of ground surface is not changed in the combination of uplift and denudation. Therefore, it is possible that the migration path and the flow velocity along the path are changed from the initial state by the depth change.

In this study, assuming disposal site composed of sedimentary rocks with uplift and denudation, we carried out integrated safety assessment where radionuclide migration along the path obtained by ground water flow analysis was evaluated taking into account the spatial and temporal changes of engineered and natural barriers. In addition, based on the results, the requirement for the investigation of natural barrier related to uplift and denudation at a candidate site was described from the regulatory point of view.

METHODOLOGY

The evolution of geological structure around a candidate disposal site with uplift and denudation causes the spatial and temporal change of ground water flow which results in the change of water composition at the site and along the migration path. We developed integrated safety assessment methodology as shown in Figure 1. It consists of ground water flow analysis, transition analysis of engineered barrier, and radionuclide migration analysis.

The ground water flow is evaluated by MIG2DF [2] which calculates the distribution of water head, and flow velocity and salt concentration by two dimensional unsteady analysis (Galerkin finite element method) coupled with advection-dispersion equation and ground water flow equation. The changes of geological formation and condition such as locational difference of uplifted height are expressed as the shape change of finite elements. In addition, the changes of the hydraulic boundary and hydraulic conductivity with denudation are taken into account in the analysis. The radionuclide migration path from the assumed repository to the outlet of natural barrier is evaluated by the trajectory analyses code, PASS-TRAC, based on the result of ground water flow analysis. It also calculates the time, length and flow velocity on the migration path which characterized with some types of rock and water composition. The transition of the boundary between rain water and salt water (salt/rain water boundary) is determined by the threshold of salt concentration which is lower limit of infiltration of ground water originated from rain water based on the results of principal component analysis for boring hole around the site. Water composition at and near the repository is determined by the relationship of the boundary and the depth of the repository

Migration parameters in engineered barrier such as diffusion coefficient and distribution coefficient in buffer material were determined based on the transition of engineered barrier obtained by MC-Buffer which was the systemized procedure using coupled mass transport and chemical reaction for each engineered barrier material such as carbon steel, bentonite clay and cement under surrounding pore water condition of excavation disturbed zone.

Finally, radionuclide migration is evaluated by one dimensional radionuclide migration analysis code, GSRW-PSA [3] which calculates the advection-dispersion equation considering of distribution equilibrium along the migration path. The code can evaluate the spatial and temporal changes of migration paths such as migration parameters and migration path itself.

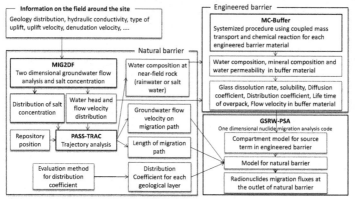

Figure 1. Outline of integrated safety assessment method for uplift and denudation

Table 1. Typical parameters for each sedimentary rock type

Geology	Distribution coefficient (m³/kg)				Porosity
	Se		Cs		
	rain	salt	rain	salt	
sedimentary rock 1	0.02	0.02	0.6	0.2	0.65
sedimentary rock 2	0.1	0.1	1.0	0.4	0.60
sedimentary rock 3	0.3	0.3	1.0	0.5	0.40

Table 2. Typical parameter for rain water and salt water

		rain water	salt water
Temperature		298 K	298 K
pH		6.8	6.3
Eh		-210 mV	-210 mV
Concentration	Na	9.74×10^{-2} mol/L	3.37×10^{-1} mol/L
	Mg	2.47×10^{-3} mol/L	6.99×10^{-3} mol/L
	Cl	9.00×10^{-2} mol/L	2.61×10^{-1} mol/L

ANALYSIS

Analytical condition

Figure 2 shows the geological condition of an assumed site which consists of three types of sedimentary rocks and a fault. Based on the survey results for an assumed site [1], the land is assumed to be uplifted by 150m for 0.5 million years (uplift rate 0.3mm/y). Orange dot line shows the ground surface at 0.5 million years in uniform uplift without denudation. Black line shows that in tilted uplift without denudation where the surface at western edge is not changed. Because denudation rate is assumed to be the same as the uplift rate, ground surface is not changed for both types of uplifts with denudation. Hydraulic conductivity is assigned for each segment at initial state and the value at three segments shallower than 700m depends on the depth. Radionuclide migration along the path for uniform or tilted uplift with or without denudation was evaluated for repository positions of P1 and P2 where vitrified waste forms are located at the initial depth of 300m from the surface, based on the trajectory analysis. Table 1 and 2 show the typical migration parameters for sedimentary rocks using for the analyses, and the characters of rain water and salt water. The outlet of the natural barrier is situated at the point

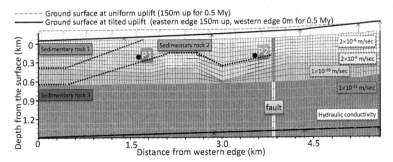

Figure 2. Geological and hydraulic conditions for ground water flow analysis with uplift

where the migration path crosses the aquifer located at 40m below the ground surface. The initial radioactive inventory in a vitrified waste form is 1.66×10^{10} Bq of Se-79 and 1.83×10^{10} Bq of Cs-135 which affect the exposure dose for long-term evaluation [4].

Results

Figure 3 shows examples of migration path by trajectory analyses in tilted uplift without denudation based on the ground water flow analysis. It shows that the hydraulic gradient increased by tilted uplift extends the migration path from both P1 and P2 after 0.1 million years. Figures 4 and 5 show the transition of migration length with type of sedimentary rocks and the

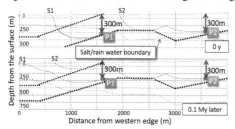

Figure 3. Example of migration path by trajectory analyses in tilted uplift without denudation

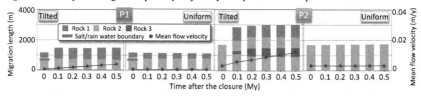

Figure 4. Transition of migration length and mean flow velocity without denudation

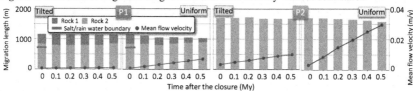

Figure 5. Transition of migration length and mean flow velocity with denudation

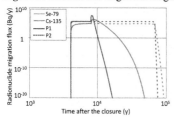

Figure 6. Radionuclide migration fluxes at the outlet of engineered barrier

mean flow velocity. Migration lengths increase over time in tilted uplift without denudation while they decrease over time in uniform uplift with denudation. The salt/ rain water boundary descends due to ground water recharge to aquifer by rain water. Salt water at P1 is replaced with rain water after 9000 years. Water composition at P2 is rain water from initial state. Figure 6 shows the calculated migration flux of Se-79 and Cs-135 at the outlet of engineered barrier, that is, the outer surface of buffer material. The fluxes start to increase at 4000 years later when confinement function of overpack is lost by uniform corrosion of carbon steel, which was determined by water composition in engineered barrier using MC-Buffer. The peak fluxes at P1 are a few orders of magnitude larger than those at P2 after 8000 years, because glass matrix is dissolved at the rate of 10^{-1} g/m^2/d after 8000 years which is a hundred times larger than that before because of supply of Mg by mineral dissolution around tunnel supporting outside buffer material. The rate depends on the Mg concentration within engineered barrier [5].

Figure 7 shows the migration fluxes at the outlet of natural barrier in the case of only uplift from P1 and P2 without denudation. Migration fluxes of Se-79 and Cs-135 at both P1 and P2 with tilted uplift are six and eight orders of magnitude larger than those with uniform uplift, respectively. This is because the migration time to the outlet for the tilted uplift is shorter due to higher flow velocity while the migration length is larger due to the increase of hydraulic gradient by tilted uplift. For only uplift, tilted uplift is more important for radionuclide migration than uniform uplift where migration distance and ground water flow velocity is not changed. It is necessary to pay attention to a possibility of the occurrence of long-term tilted uplift. In particular, it is necessary to investigate the fold structure around a candidate site carefully.

Figure 8 shows the migration fluxes at the outlet of natural barrier in the case of both types of uplift with denudation. Migration fluxes at both P1 and P2 with uniform uplift are larger than those with tilted uplift, because migration length and time at uniform uplift are shorter than those at tilted uplift where the depth is larger than that in uniform uplift. In particular, the difference of

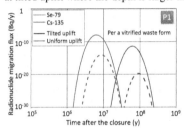 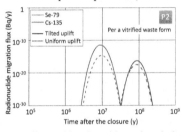

Figure 7. Radionuclide migration fluxes at the outlet of natural barrier without denudation

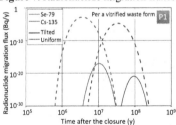 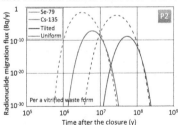

Figure 8. Radionuclide migration fluxes at the outlet of natural barrier with denudation

the peak fluxes of both radionuclides are larger than ten orders of magnitude, because mean flow velocity is seven times larger than that in tilted uplift in addition to shorter length of migration path of sedimentary rock 1 where the distribution coefficient is smaller. The peak fluxes at P2 are approximately ten orders of magnitude larger than those at P1 in tilted uplift, because the mean flow velocity after 0.5 million years along the path from P2 with 200m depth is more than ten times larger than that from P1 with 255m depth. The hydraulic conductivity increases with decreasing depth from the ground surface because of the release of rock pressure. Therefore, radionuclide migration at a shallower depth is accelerated by the increase of flow velocity. In the case of combination of uplift and denudation, uniform uplift is more important than tilted uplift because migration length and time decreases in uniform uplift where the depth decreases more than that in tilted uplift. The combination of uniform uplift and denudation has the largest effect on radionuclides migration fluxes in the cases. Therefore, it is necessary to validate the denudation model with consideration of long-term spatial and temporal changes around a candidate site. It is also important to investigate the structures of geological layers including uplift and denudation, the hydraulic conditions, and the recharge and outlet of the ground water around a candidate site, and to determine the appropriate place, depth and layout design of the repository based on the analytic results of radionuclide migration along the paths.

CONCLUSIONS

Integrated safety assessment methodology that analyzes radionuclide migration reflecting the spatial and temporal changes of disposal systems was developed for a geological disposal site with uplift and denudation, and then some case analyses for an assumed site were carried out. The combination of uniform uplift and denudation has the largest effect on the radionuclide migration because the ground water flow velocity increases with decreasing the depth from the ground surface. In the case without denudation, tilted uplift has larger effect than uniform uplift because flow velocity in tilted uplift increase with increasing hydraulic gradient. The long-term changes of geological structures including the uplift and denudation, the hydraulic conditions, and the recharge and outlet of the ground water around a candidate site should be carefully investigated to determine the appropriate place, depth and layout of the repository.

ACKNOWLEDGMENTS

This research is funded by the Secretariat of Nuclear Regulation Authority, Nuclear Regulation Authority, Japan.

REFERENCES

1. Niizato T., Yasue K., A study on the long-term stability of the geological environments in and around the Horonobe area. *J Nucl Fuel Cycle and Env*, **11** (2), 125-137, March (2005). [in Japanese]
2. Kimura H., The MIG2DF Computer Code User's Manual, JAERI-M 92-115, (1992).
3. Takeda S. *et. al.*, Assessment on Long-term Safety for Geological Disposal of High Level Radioactive Waste, JAEA-Research 2009-034, (2009). [in Japanese]
4. JNC., 2nd Progress Report on Research and Development for the Geological Disposal of HLW in Japan, JNC Technical Report TN1410 2000-001, Japan Nuclear Cycle Development Institute, Tokai-mura Japan, (2000). [in Japanese]
5. Maeda T. et. al., Corrosion Behavior of Simulated HLW Glass in the Presence of Magnesium Ion, International Journal of Corrosion, Volume 2011 (2011)

Mater. Res. Soc. Symp. Proc. Vol. 1744 © 2015 Materials Research Society
DOI: 10.1557/opl.2015.335

Pore and mineral structure of rock using nano-tomographic imaging

Jukka Kuva[1], Mikko Voutilainen[2], Antero Lindberg[3], Joni Parkkonen[1], Marja Siitari-Kauppi[2] and Jussi Timonen[1]

[1]University of Jyväskylä, Department of Physics, P.O. Box 35, 40014 University of Jyväskylä, Finland
[2]University of Helsinki, Department of Chemistry, P.O. Box 55, 00014 University of Helsinki, Finland
[3]Geological Survey of Finland, Betonimiehenkuja 4, 02151 Espoo, Finland

ABSTRACT

In order to better understand the micrometer-scale structure of rock and its transport properties which arise from it, seven monomineral samples from two sites (Olkiluoto and Sievi, Finland) were studied with micro- and nanotomography and scanning electron microscopy. From the veined gneiss of Olkiluoto we studied biotite, potassium feldspar, plagioclase (composition of oligoclase) and cordierite, and from Sievi tonalite biotite and two grains of plagioclase (albite). These minerals were the main minerals of these samples. Samples were carefully separated and selected using heavy liquid separation and stereomicroscopy, their three dimensional structure was imaged using X-ray tomography, and their precise mineral composition was determined using scanning electron microscopy and energy-dispersive X-ray spectroscopy (SEM/EDS). The micrometer-scale mineral structure of these samples was observed together with their pores and fissures, and alteration effects were identified whenever applicable. Nanotomography combined with SEM analysis was concluded to be a good tool for analyzing effects of alteration in monomineral samples.

INTRODUCTION

In Finland the spent nuclear fuel will be deposited deep in the crystalline host rock. The safety analysis of spent nuclear fuel deposition includes a release scenario, where radionuclides migrate in the rock matrix. One of the major retarding factors for migrating radionuclides in crystalline rock is diffusion in the pores (fissures, inter- and intragranular pores) [1-3]. Typically, in crystalline rock, the majority of pore apertures are in a nanometer-scale and they have a significant effect on the diffusion of radionuclides. Sizes of these apertures are also closely linked to the surface area available for sorption of radionuclides, which is another important retarding factor. We set out to study the microstructure of rock in this scale from monomineral grains. Another aim was to see if X-ray tomography can be used to study in a micrometer scale the effects of alteration [4] in the structure of rock, and to this end two very different rock samples were chosen: slightly altered veined gneiss from Olkiluoto and heavily altered tonalitic igneous rock from Sievi.

X-ray tomography has already been widely used as a tool to study the 3D structure of geological samples, and has proved to be very useful for this purpose [5-7]. So far there have, however, been much less studies applying nano-tomography on geological samples [8], so we also set out to see how much additional information it can provide. Identification of minerals cannot however be performed by X-ray tomography. To this end, X-ray tomography and SEM/EDS have successfully been combined in previous studies [9].

SAMPLES AND EXPERIMENTAL DETAILS

The mineralogy and structure of the intact rock samples was determined from previously manufactured thin sections [10, 11]. The Olkiluoto sample was veined gneiss with biotite, quartz, plagioclase, potassium feldspar and cordierite as the main minerals. This sample was small grained (0.3 – 1 mm) and only slightly altered. The medium grained (2 – 5 mm) Sievi sample was heavily altered. Main minerals were plagioclase, biotite and different opaque minerals. Both biotite and plagioclase were heavily altered to chlorite and albite. There was almost no quartz.

Rock samples were crushed and sieved; then the different fractions were separated with heavy liquids [12]. Finally, so as to get as small and pure mineral grains as possible, individual grains were hand-picked using a stereo microscope.

Eventually seven minerals were imaged with X-ray tomography: plagioclase, potassium feldspar, biotite and cordierite from the Olkiluoto rock and biotite (chlorite) and two grains of plagioclase (albite) from the Sievi rock. It was assumed that comparing plagioclase and biotite grains of these totally different rock samples, at least some alteration induced changes could be observed in their mineral and pore structure. Samples were imaged with micro- and nano-tomography at the University of Jyväskylä. Nano-tomographic imaging was done using an XRadia NanoXCT with a 65 nm voxel size and a 65 μm field of view. This device has a fairly monochromatic X-ray beam of 8.0 keV. According to the manufacturer, the real spatial resolution with these settings was 150 nm. Microtomography was done to help to choose the right parts of the samples for nano-tomography, and to generally get a larger field of view. It was done using an XRadia MicroXCT 400 with a 0.3 μm voxel size and a 600 μm field of view. This device has a polychromatic X-ray beam with maximum intensity at 8.4 keV.

Finally, the precise elemental composition, which confirmed the minerals of the samples, was done using a scanning electron microscope and energy-dispersive X-ray analysis (SEM/EDS and INCA D point program) using 20 keV energy at the Geological Survey of Finland.

RESULTS

Four feldspar samples are shown in figure 1: the potassium feldspar and plagioclase grains from Olkiluoto and two plagioclase grains from Sievi. According to SEM/EDS analyses shown in table I, both plagioclases from Sievi were albite with no calcium in the structure, the Olkiluoto plagioclase was oligoclase and the Olkiluoto feldspar was potassium feldspar. Pores can be observed in the less dense parts of the Olkiluoto potassium feldspar grain and there are two main components visible. The first Sievi plagioclase sample shows signs of extensive hydrothermal alteration: there are more pores and the mineral structure is more heterogeneous in comparison with the less altered Olkiluoto feldspars. Pores in the first Sievi plagioclase sample were significantly larger, with apertures commonly up to 15 μm, whereas the biggest apertures in the Olkiluoto potassium feldspar sample were quite rare and only 6 μm. Some of these samples had plenty of pores, but they did not show visible connection with the used resolution. The Olkiluoto plagioclase (oligoclase) sample and the second Sievi plagioclase (albite) sample had a homogeneous mineral structure, showing only one main component and no pores within the resolution limit of the technique. The mineralogical study of a thin section of the Sievi sample had proved that part of the plagioclase had recrystallized [9]. The second Sievi plagioclase sample for this tomography study may be one of those grains.

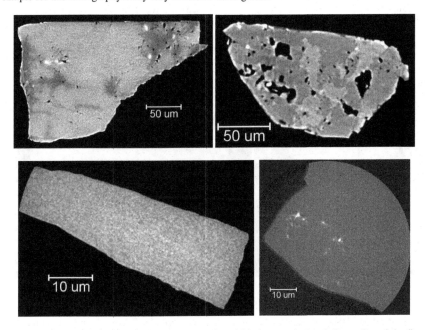

Figure 1. Cross sections of the Olkiluoto potassium feldspar sample (top left panel) and the first Sievi plagioclase (albite) sample (top right panel) as determined by X-ray microtomography, with two main mineral components and several pores visible in both, and those of the Olkiluoto plagioclase (oligoclase) sample (bottom left panel) and the second Sievi plagioclase (albite) sample (bottom right panel) as determined by X-ray nano-tomography, with no pores and only one main component visible.

Table I. SEM/EDS results for four feldspar samples, two from Olkiluoto and two from Sievi.

Sample	Na %	Al %	Si %	K %	Ca %	Identification
Sievi plagioclase #1	11.79	20.66	66.97	0.58	0	Albite
Sievi plagioclase #2	8.26	21.41	70.34	0	0	Albite
Olkiluoto plagioclase	10.19	21.31	65.54	0.50	2.46	Plagioclase (oligoclase)
Olkiluoto feldspar	2.25	15.86	62.19	19.7	0	Potassium feldspar

Cross sectional images of biotite grains from Olkiluoto and Sievi are shown in figure 2. According to SEM/EDS analyses shown in table II, the Olkiluoto biotite was unaltered and contained a small amount of titanium and manganese. Tiny grains of uranium were also found in the sample. The Sievi biotite was strongly altered to chlorite, which was seen as the decrease of potassium and the increase of magnesium content in comparison with an unaltered biotite. In both samples the lamellar structure and some of the interlamellar pores were visible in the tomographic images. However, in the Sievi chlorite the grain space between the lamellas was filled with heavier elements. These bright areas witnessed the infiltration of a heavier mineral, probably sulfide or titanium oxide, during alteration. These observations are indicative of heavy alteration. It has to be emphasized that the largest lamellar pores have probably opened and also other types of structural change is likely to have occurred (see the curved lamella in the Olkiluoto sample in figure 2, left panel) during sample preparation.

Figure 2. Cross sections of the Olkiluoto biotite sample (left panel) and the Sievi altered biotite (chlorite) sample (right panel) as determined by X-ray nanotomography.

Table II. SEM/EDS results for two biotite samples, one from Olkiluoto (OL) and one from Sievi.

Sample	Mg %	Al %	Si %	K %	Ca %	Ti %	Fe %	Mn %	Identification
OL biotite	11.33	14.68	40.78	9.25	0	3.41	19.57	0.98	Biotite
Sievi biotite	16.93	20.05	40.26	3.24	0.87	1.47	17.19	0	Chlorite

The last mineral grain analysed by X-ray nano-tomography and SEM/EDS (see table III) was the Olkiluoto cordierite sample. In the cross sections of the tomographic image three different gray values were observed (see figure 3). It was concluded that the dark gray areas were unaltered cordierite and the light gray veinlets were composed of an alteration product of cordierite. These veinlets contained a clayish mixture called pinite, and the whole grain could be classified as moderately altered. The roundish mineral in the upper part of the grain was a quartz inclusion typical of cordierite, which could be identified from its shape. The white areas might be minerals which contained phosphorus, and more iron and magnesium than other parts of the grain, as suggested by the SEM analysis.

Table III. SEM/EDS results for the Olkiluoto cordierite sample.

Sample	Na %	Mg %	Al %	Si %	K %	Ca %	Fe %	Identification
Olkiluoto cordierite	2.95	2.34	34.68	53.07	1.62	0.83	4.51	Cordierite

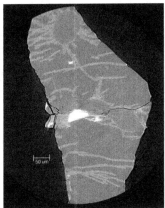

Figure 3. A cross section of the Olkiluoto cordierite sample as determined by X-ray microtomography.

CONCLUSIONS

Using X-ray tomography we were able to see the microstructure of different monomineral samples, and to observe signs of alteration. When tomography was combined with SEM/EDS analysis, we were able to investigate more precisely the changes caused by alteration in the elemental composition of minerals. SEM/EDS was used for identification of the minerals themselves as well as of the different compounds in them.

As expected, the Sievi samples were more altered than the corresponding Olkiluoto samples. We were able to observe the pores inside the minerals, especially in the albite from Sievi, where they were significantly larger and more numerous than in the Olkiluoto feldspars. Lamellar structure was observed in both Sievi and Olkiluoto biotites by tomography. SEM/EDS confirmed that the biotite from Sievi was fully altered to chlorite, whereas the Olkiluoto biotite was unaltered. The Olkiluoto cordierite was found to be moderately altered.

We can conclude that nano-tomography combined with SEM/EDS is a good tool for analyzing the effects of alteration in monomineral samples, as well as for investigating their inner structure. These methods can give extensive information about the pore and mineral structure in a nanometer scale, as well as about changes caused by alteration. It is noteworthy that tomographic imaging has to be assisted by petrographic analysis tools.

We are now conducting an experiment in which rock samples are imaged with tomography before and after being immersed for several months in a supersaturated cesium chloride solution. The aim is to enhance the X-ray absorption coefficient in the places, where cesium has intruded the rock, so as to have more information about the micro- and nanometer-scale structure of rock and minerals, as well as about connected pore network and the diffusion of cesium in these samples.

REFERENCES

1. I. Neretnieks, J. Geophys. Res. **85**(B8), 4379-4397 (1980)
2. M. Voutilainen, P. Kekäläinen, A. Hautojärvi and J. Timonen, Phys. Chem. Earth **35**(6-8), 259-264 (2010)
3. P. Kekäläinen, M. Voutilainen, A. Poteri, P. Hölttä, A. Hautojärvi and J. Timonen, Transport Porous Med. **87**, 125-149 (2011)
4. A. Rossi and R. Graham, Soil Sci. Soc. Am. J. **74**, 172-185 (2010)
5. V. Cnudde and M.N. Boone, Earth-Sci. Rev. **123**, 1-17 (2013)
6. F. Fusseis, X. Xiao, C. Schrank and F. De Carlo, J. Struct. Geol. **65**, 1-16 (2014)
7. J. Kuva, M. Siitari-Kauppi, A. Lindberg, I. Aaltonen, T. Turpeinen, M. Myllys and J. Timonen, Eng. Geol. **139-140**, 28-37 (2012)
8. S. Bugani, M. Camaiti, L. Morselli, E. Van de Casteele and K. Janssens, X-Ray Spectrom. **36**, 316-320 (2007)
9. M. Voutilainen, M. Siitari-Kauppi, P. Sardini, A. Lindberg and J. Timonen, J. Geophys. Res. **117**, B01201 (2012)
10. J. Sammaljärvi, A. Lindberg, J. Ikonen, M. Voutilainen, M. Siitari-Kauppi and L. Koskinen in *Scientific Basis for Nuclear Waste Management XXVII*, edited by L. Duro, J. Giménez, I. Casas and J. de Pablo, (Mater. Res. Soc. Symp. Proc. **1665**, Pittsburgh, PA, 2014) pp. 31-37
11. A. Lindberg and M. Paananen, Teollisuuden Voima Oy Work Rep. **92-34** (1992)
12. R.P. Bagioni, Environ. Sci. Technol. **9**(3), 262-263 (1975)

AUTHOR INDEX

Ojovan, Michael I., 67, 153

Parkkonen, Joni, 235
Parruzot, Benjamin, 139
Puranen, A., 217

Qafoku, Nikolla P., 43, 139

Reiser, Joelle, 101, 139
Remizov, Michael B., 73, 79
Rossignol, Sylvie, 15
Roth, O., 217
Ryan, Joseph, 139

Sakuragi, Tomofumi, 3, 21
Schumacher, Stéphan, 127
Serne, R. Jeffrey, 43
Shimada, Taro, 229
Shiryaev, Andrey A., 73, 79

Siitari-Kauppi, Marja, 235
Spahiu, K., 217
Stefanovsky, Sergey V., 73, 79
Stennett, Martin C., 185

Takeda, Seiji, 229
Tan, Shengheng, 67
Tanaka, Tadao, 229
Timonen, Jussi, 235
Travis, Karl P., 53, 193, 205

von Schenck, Henrik, 223
Voutilainen, Mikko, 235

Wall, Nathalie A., 93, 101, 139
Weaver, Jamie L., 93, 101, 139
Williams, Christopher D., 53

Yoshida, Satoshi, 21

SUBJECT INDEX

Printed in the United States
by Baker & Taylor Publisher Services